P.G. HA
31/5/50

The story of RUDGE motorcycles

In the same series

Norton Story
2nd edition
Bob Holliday

The Story of Triumph Motor Cycles
4th edition
by Harry Louis and Bob Currie

The Story of BMW Motor Cycles
by Robert Croucher

The Story of BSA Motor Cycles
by Bob Holliday

The Story of Royal Enfield Motor Cycles
by Peter Hartley

The Story of Panther Motor Cycles
by Barry M. Jones

In preparation

The Story of AJS Motor Cycles
2nd edition
by Bob Holliday

The Story of Velocette Motor Cycles
by Peter Carrick

The story of RUDGE motorcycles

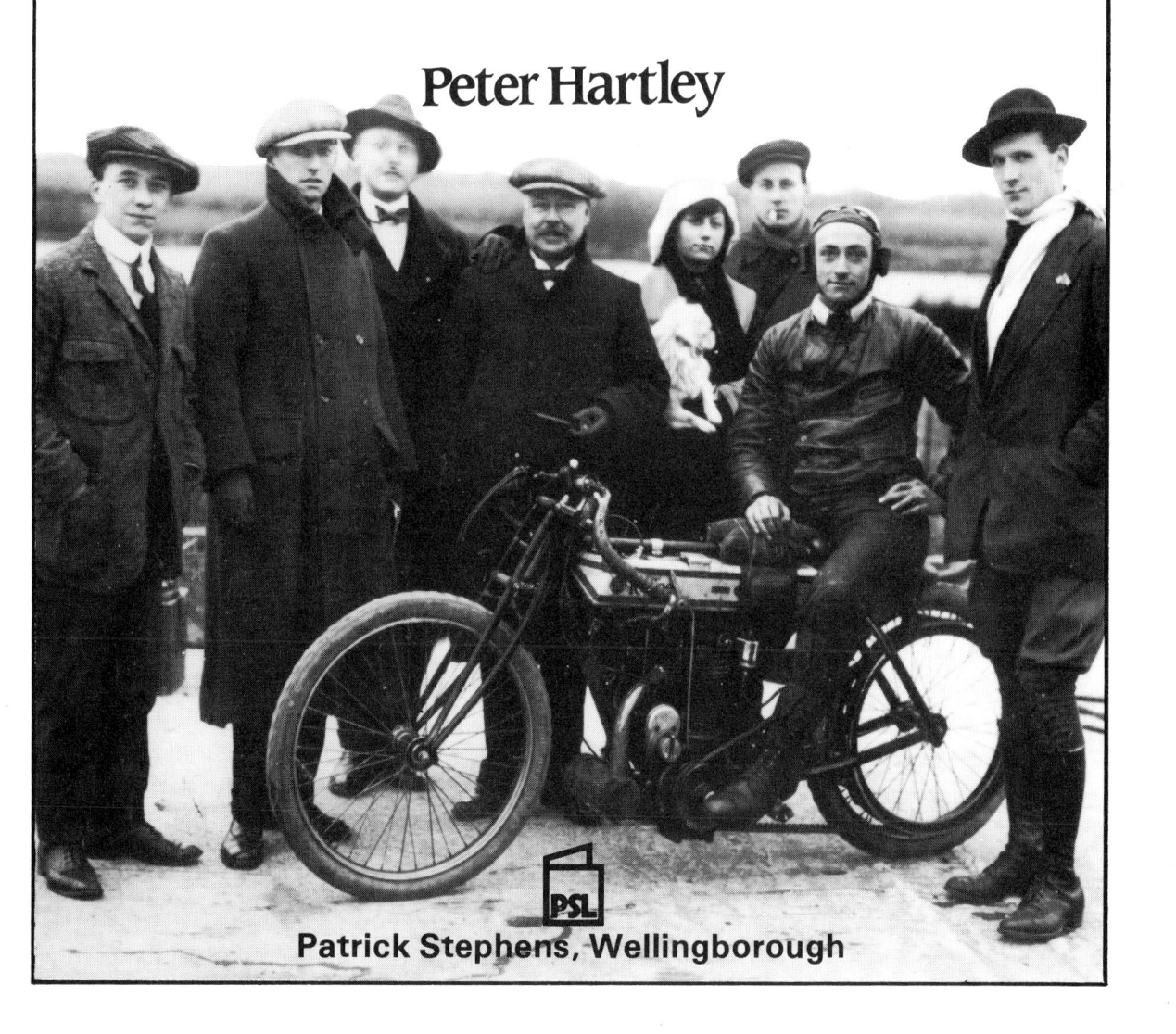

Peter Hartley

Patrick Stephens, Wellingborough

© Peter Hartley 1985

All rights reserved. No part of this publication may be reproduced, stored in a retrieval system, or transmitted, in any form or by any means, electronic, mechanical, photocopying, recording or otherwise, without prior permission in writing from Patrick Stephens Limited.

First published in 1985

British Library Cataloguing in Publication Data

Hartley, Peter, *1933-*
The story of Rudge motorcycles.
1. Rudge motorcycle—History
1. Title
629.2'275   TL448.R82

ISBN 0-85059-511-8

Patrick Stephens Limited is part of the Thorsons Publishing Group

Photoset in 10 on 11 pt Times by MJL Limited, Hitchin, Herts. Printed in Great Britain on 120 gsm Champagne coated cartridge and bound by William Clowes Limited, Beccles, Suffolk, for the publishers Patrick Stephens Limited, Denington Estate, Wellingborough, Northants, NN8 2QD, England.

Contents

Preface *6*
1 The Wizard from Wolverhampton (1860-1908) *7*
2 Power and pedal (1909-1911) *15*
3 Gearing for success (1912-1913) *28*
4 Experiments, prototypes and winners (1914) *42*
5 Rudges at war (1915-1919) *48*
6 A change of gear (1920-1922) *51*
7 Four speeds and four valves (1923-1924) *56*
8 Back to racing (1925-1927) *60*
9 Rewards from racing (1928) *66*
10 Things start to go right (1929) *71*
11 TT triumphs (1930) *79*
12 Hard times (1931-1932) *90*
13 The Syndicate (1933-1935) *104*
14 The end of the road (1936-1943) *112*
15 Old Rudges never die (1946-1985) *121*

Appendices
1 Some representative Rudge-Whitworth valve and ignition timings *123*
2 The last frame numbers in any given year on Rudge-Whitworth motorcycles *124*
Index *126*

Preface

Many of the technical features of modern motorcycle design, taken for granted by today's generation of motorcyclists, were pioneered *not* by the Japanese but by *British* motorcycle manufacturers. Rudge-Whitworth Ltd was one of the leaders in this respect, with developments ranging from four-valve engines in production machines to the use of megaphone exhausts in racing. Even earlier, the firm had pioneered the first really successful infinitely variable speed transmission which was later to appear in a modified form on the Dutch DAF car. The motorcycle that made use of it, the Rudge-Multi, was so successful that it remained in production for almost 12 years.

The Rudge name for technical innovation and excellence began during the life of the firm's founder Dan Rudge, and was maintained by George Woodcock, John Pugh and lastly George Hack, until the firm eventually ceased production in 1940. Technical excellence was the theme all through the company's history.

This frequently meant being first with completely new and untried ideas later copied by rival manufacturers. Being first with a new design also meant being first to be faced with all the problems of making it a practical proposition and this proved expensive. Development took the form not only of work in the test shop but also of fielding a team of racing machines during the late 1920s and early 1930s, at a time when everyone else was retrenching, just prior to entering the throes of one of the most savage economic depressions in the history of the western industrialised world. To recover its development and racing costs, the company was forced to start raising its prices at a time when most other motorcycle manufacturers were attempting to reduce theirs in an effort to drum up sales. This was hardly a way to stay successfully in business and by 1933 the firm was in receivership.

That might have been the end of the tale, but then the company was bought by one of its largest creditors, EMI, the electrical and electronics concern, and things were all set for a revival. Then war came and through an unexpected turn of fate — the company was never very lucky — production was stopped in 1940, never to be restarted. That was the end of production of a very fine range of motorcycles.

Why then did I decide to write this tale, the story of a make of machine that ended before I was old enough to know the difference between a cylinder head and a spanner; the story of a marque that ceased to be produced more than 40 years ago? Well, it is a marque that is by no means forgotten. There exists a flourishing Rudge Enthusiasts' Club and there are still a large number of Rudge-Whitworth motorcycles alive and well and run regularly on the roads of Britain in the hands of enthusiasts. I also think it is a story well worth telling. On the surface it is a story of apparent failure in the face of insuperable odds; but is it? True, there are no longer any Rudge-Whitworth machines being made, but many of their design features are used widely in modern motorcycles, not least by the Japanese. The net result has therefore been of direct benefit to the motorcyclists of the 1980s, even though many of them have been blissfully unaware of the British origins of many of their machines' design features. Messrs Rudge, Woodcock and Pugh would, I suspect, not have been too displeased with the final products, even though they might have wished them to have been British made.

Peter Hartley,
Borstal, Kent, 1984

The Wizard from Wolverhampton (1860-1908)

When Dan Rudge came back to his home town of Wolverhampton after serving in the Crimean War with the British Army's 38th Regiment of Foot, he did what many an old soldier before him had dreamt of doing, he opened up his own pub, the 'Tiger's Head' in Church Street. Meanwhile his old ex-army comrade, Henry Clarke, had started a wagon-wheel building business nearby, called the Temple Street Wheel Works.

A born mechanic, Dan spent much of his spare time in Henry's works tinkering with carts and wagons. There was not a man in Wolverhampton, it was said, who could mend a wheel as quickly or as neatly as Dan Rudge, despite the fact that he was not a 'professional' at the game. He greased the axles of the customers' carts, regulated their brakes and carried out any other maintenance on them he deemed necessary.

In 1868, the Coventry Machinists Company had started mass producing an improved version of velocipede or 'boneshaker'. By 1870, James Starley, formerly in change of design and production with this company, had set up his own business in Coventry to produce his Ariel bicycles of the 'high' or 'ordinary' (penny-farthing) type and within a month or so this news had become common knowledge throughout the country. Already (in 1868) Henry Clarke had launched another business enterprise, the Cogent Cycle Company, to manufacture bicycles in premises in Darlington Street, Wolverhampton. This was Henry's second venture into the world of cycles, having built boneshakers at an even earlier date. With all this cycling activity going on and with his mechanical bent, it was not surprising therefore that Dan Rudge should also take an active interest in two-wheelers at this time.

Dan Rudge, founder of the Rudge concern in Wolverhampton.

The type of 'high bicycle' produced by Dan Rudge and Henry Clarke in the 1870s.

The year 1869 saw two other Wolverhampton men interested in the new pedal cycles, Walter Phillips and George Price. Phillips had also invented an improved form of velocipede. Price, on the other hand, was more interested in the business possibilities of cycle-making. They decided to go into business together, but Phillips, the more mechanically minded of the two, realised that successful production of a two-wheeler would need a good, skilled mechanic to undertake the project and sort out the problems. With this in mind, early in 1869 the two partners approached Dan with the business proposition of manufacturing the Phillips' design of velocipede. Dan soon became very interested and settled down to serious production of bicycles in collaboration with Henry Clarke, who supplied the wheels.

It was at about this time that a Frenchman, who had met Henry Clarke in the Crimea, called on him riding a type of French velocipede. Dan and Henry were amazed by the apparent ease with which the machine sped along. Determined to unravel the mystery of how this was achieved, they took the Frenchman into the 'Tiger's Head' and plied him with strong drink until he was completely inebriated. When he passed out, the two conspirators stripped down his machine completely and discovered its secret — the hubs ran on ball, not plain, bearings. They quickly put everything back together again and their French friend went on his way again none the wiser. It was not long after this episode that Dan Rudge took out patents of his own, covering ball-bearing wheel hubs.

Dan tested the first machines he produced and, at the first-ever cycle race meeting to be held in Wolverhampton, at the Molineux Gardens, in the summer of 1869, he took one to victory in one of the events.

By the end of 1870, Dan had produced a small number of high-quality machines of the 'ordinary' 'penny-farthing' type, with an enormous front wheel over which the rider sat, whilst turning it by means of directly attached pedals.

Dan did not restrict his copying of good designs to just hub bearings for, three months after James Starley had advertised his Ariel 'high' bicycle in *The English Mechanic* magazine, *The Times* newspaper for January 2 1872 reported that: 'We have now been informed by telegraph that Clarke and Rudge in Wolverhampton are copying this machine'.

By 1873, the fledgling firm of Dan Rudge, operating from a shed near the 'Tiger's Head', had established a firm reputation as a maker of high quality bicycles. Unhappy with the poor standards of design and construction then prevalent, Dan Rudge initiated a number of innovations. That year he patented his famous adjustable bearing (British Patent 520) designed to reduce friction between bearing surfaces. This he fitted to his cycles, thereby improving their efficiency and ease of riding. His machines soon gained a reputation for fast starting in races and their popularity rocketed.

As a result of the success of these machines, in many of the cycle races of the day, riders of Rudge-built machines were frequently obliged to concede a 20- or 30-yard (18- or 27-m) start to other cycle makes not fitted with these bearings! They were considered that much superior to their competitors in terms of all-round efficiency, a point not lost on the buying public! This unintentional good publicity further increased the demand for Rudge bicycles. Dan Rudge's cycle business went from strength to strength, but the combined effects of burning the candle at both ends, running both his cycling and inn-keeping businesses plus his cycle-racing activities, were affecting his health detrimentally. He had had a number of illnesses over the years as a result, but still would not ease up the frantic pace of his life and in the early summer of

George Woodcock, the retired solicitor and entrepreneur, who in 1879 started building up his cycle-making empire in Coventry.

1880 fell ill for the last time, dying on June 26 1880 only a short time after taking out a patent (British Patent 3924) on a small-wheeled safety cycle of modern design, one of the first of its kind but which unfortunately, as a result of his death, was still-born.

With the death of this remarkable man, his business interests fell into the hands of his wife and sales continued undiminished for a few months, but of course no new designs emerged from the Rudge stable, Dan's son Harry Rudge then being too young to take an interest in the business. It was at this stage that another extraordinary man came into the Rudge picture — George Woodcock — an entrepreneur of genius.

Born in 1836, Woodcock started his business career not in engineering but the law, his father being a well known and trusted Coventry solicitor and the senior partner of the long-established firm of Woodcock & Twist of Coventry. George Woodcock started in the family legal concern early in his career and ultimately became, like his father before him, a highly successful solicitor. In fact, so prosperous did he become in his profession that in about 1874 he was able to retire with a sizeable fortune at his disposal. One of nature's restless spirits, though, having decided to leave the legal profession, he could not bear to remain inactive for the remainder of his life. He became immersed in various business matters and in particular started to take an interest in the infant cycle trade. In 1879 he converted this interest into positive action. He bought up a number of small cycle concerns in and around Coventry, including the firms of Haynes & Jefferis which produced the Coventry Rotary Tricycle and had works in Spon Street and Crow Lane, and Smith, Starley & Co Ltd which produced sewing machines to the original James Starley designs. This last purchase allowed James Starley to concentrate on his sons' newly launched cycle concern — Starley Brothers.

After acquiring these concerns, George Woodcock set about rationalising their assets. He sold off the old Haynes & Jefferis' Spon Street works and set about fitting out each floor of the Crow Lane factory with the very latest production machinery. The reorganised firm was re-launched under the name of The Tangent & Coventry Tricycle Co Ltd and set about the production of five models. They comprised three bicycles of the ordinary or penny-farthing type (the Special Tangent Bicycle, the Tangent Bicycle and the Swiftsure) and two tricycles (the Coventry Lever Tricycle and the Coventry Rotary Tricycle). The two tricycles were inherited from the former firm of Haynes & Jefferis. First produced at the beginning of 1877, the Coventry Lever Tricycle was originally designed and patented by James Starley. It had a large driving wheel on the left of the rider, who drove it by means of two levers whilst seated on a crossbar in the centre of the machine. Steering was by means of a tiller controlling two small wheels located one behind the other, in the same track, on the right-hand side of the machine.

This tricycle was originally developed by James Starley from his unsuccessful ladies' machine. It was subsequently given a rotary chain drive action and re-christened the Coventry Rotary Tricycle by Harry Lawson who had patented its design when working for Haynes & Jefferis in 1877 (British Patent 972). This machine was fitted with tangent spokes, another idea to come from James Starley's fertile brain.

To return to the late Dan Rudge's business affairs: after his death in 1880, his ex-colleague, the much respected cycle-maker, Walter Phillips, acting as go-between, arranged the purchase of Dan's enterprise from his widow by George Woodcock. The agreement was that she should receive a cash sum plus a regular pension. This purchase included the acquisition by Woodcock of the Rudge adjustable ball-bearing patent and the services of some of his highly skilled former employees. Following this merger, a further name change occurred and The Tangent & Coventry Tricycle Co Ltd was relaunched in Coventry under the new and rather clumsy name of D. Rudge & The Coventry Tricycle Company, which was chosen to take advantage of all the business goodwill associated with the Rudge name.

One of George Woodcock's greatest assets was his ability to size up people and opportunities accurately, and he saw in Harry Lawson a keen business brain. He appointed Harry as Manager of the new concern. It was then (in 1880) that Lawson brought out his chain-driven safety bicycle, the 'Bicyclette', which had been patented the previous year (British Patent 3934, September 30 1879).

Compared with the then conventional penny-farthing or ordinary cycles, it would be termed a 'penny-ha'penny', since it had a back wheel which, although small, was only half the diameter of the front wheel. The rear wheel was driven by a chain from a crank mounted between the wheels.

This machine had what was known as 'bridle-rod' steering. Lawson's patent claimed that 'the steering handle. . . is connected to the said fork by means of a bar, *or the last mentioned fork is arranged at an angle so as to bring the handle within reach of the rider*'. It is these words in italics that are the crux of Lawson's new design, in that he had invented the raked steering-head cycle frame with its natural stability.

The machine was a major advance on the ordinary high cycle and, to commemorate its invention, George Woodcock presented Harry Lawson with a gold watch engraved with the legend 'To the inventor of the Safety Bicycle'. Unfortunately, despite this and being well publicised at the Third Stanley Show staged in 1880 at the Holborn Town Hall, London, it was not a financial success. It was not until 15 years later that Lawson was to be publicly acclaimed as the inventor of the rear-driven safety cycle. This was due to several factors. Firstly, the cycle-buying public was fairly conservative in its tastes and did not take readily to radically new designs. Rival cycle makers tended to ridicule it and nicknamed it the 'Crocodile'. Strangely enough, its major contribution has been to the French language in that, because the Rudge marketing organisation was then (in 1880) setting up palatial showrooms in Paris where the machine was put on display, the French soon started using the name 'bicyclette' to describe all safety bicycles and the word has stuck to this day. Before long, large numbers of the now famous Rudge cycles, of all types, were being shipped over to the Continent at regular intervals. In this country the Bicyclette, however, was known as the 'Woodcock'.

The major reason for the commercial failure of the Bicyclette, though, was a technical one. With such a small rear wheel, an unduly large sprocket (and cranks) had to be employed for a reasonable gear ratio to be obtained. This meant that the cranks restricted the turning angle of the front wheel, as George Woodcock found to his discomfort when he fell off the machine during a test run. He was not pleased and let Lawson know the fact in no uncertain terms. It is rumoured that this incident, compounded by the cost of stopping production of the machine due to its poor sales, started the eventual rift that developed between the two men. That, however, was still some years away.

It was at around this time that Harry Lawson patented his first motorcycle, which was propelled by 'compressed gas' according to the specification (British Patent 3913, dated September 20 1880). Nothing seems to have come of this commercially and his next involvement with a powered vehicle, the notorious Pennington motorcycle of the late 1890s, suggests that it would probably not have been a great success anyway.

With the formation of D. Rudge & The Coventry Tricycle Company as a result of the amalgamation of the various companies acquired by George Woodcock, a serious review of its entire range of machines was undertaken. The new company already owned a shop in Victoria Street, London, which was to help greatly in promoting the sales of Rudge bicycles. Meanwhile, the Coventry Rotary Tricycle had been improved by the fitting of cranks, pedals and a bottom bracket giving a *central* chain drive to a sprocket mounted on an extension of the rear wheel spindle. This wheel was geared up to about 56 in (1.44 m), which was the distance moved by the machine for one revolution of the pedal cranks.

Rechristened the 'Rudge Rotary', this tricycle gained great popularity and considerable numbers were sold. It was ridden in many races by Jack Morley of Manchester and that great racing cyclist of the day, Matthew 'Jumbo' Lowndes of Congleton, Cheshire, who was beaten only once, by Alfred Nixon, in an epic 100-mile match race at the Crystal Palace Track in South London. After the race Lowndes demonstrated the controllability of his machine by riding it down the steps to the platform at the Crystal Palace low-level railway station.

The first models of the Rudge Rotary had a so-called 'garden seat', but later models gave the purchaser the option of a seat or a saddle. Ladies, of whom quite a number rode tricycles at this time, generally preferred seats to saddles. Her Royal Highness the Princess of Wales, later to become Queen Alexandra, regularly used a Rudge Rotary tricycle, having had a machine specially built with pneumatic tyres long after the type had become obsolete.

The Rudge 'ordinary' high machines became very popular for racing, being of high quality and weighing a mere 21½ lb (9.8 kg). In fact both Harry Lawson and Richard Howell gained world championships riding these machines. So popular were they that the less successful Bicyclette was withdrawn in 1881 to allow concentration of production on the high models.

While sales were burgeoning at home and abroad between 1881 and 1884, further thought was being given to an improved version of the Bicyclette. After much testing of a new prototype in and around Coventry, the improved Bicyclette was launched in 1885 and started to sell well. The same year George Woodcock decided to increase his capital for expansion by turning the firm into a private limited company, supported by subscriptions from 26 of his business friends in the Birmingham/Coventry area. The old clumsy name was dropped and the new enterprise was christened D. Rudge & Co Ltd. It was exports to the USA, France, Germany and, for the first time, to Denmark, however, which largely accounted for the rapid up-turn in the firm's fortunes that made this expansion possible. The Victoria Street venture had proved very successful in promoting sales, so the company had set up a depot system throughout the whole of Britain to sell direct to the customer. It was to prove highly successful, albeit creating considerable animosity between Woodcock and the rest of the cycle trade.

A number of publicity stunts were tried around this time, including the construction of a cycle boat using standard Rudge cycle parts. Tested on a local Coventry canal, it was designed to be propelled by three riders and, on its first trial run, Bob Walker Smith, Selwyn F. Edge and Frank Cooper provided the motive power. All three were eventually to achieve fame in either the cycling, motorcycling or car worlds.

The firm also produced at this time an unusual tricycle or 'road sculler' for the keen oarsman to practise upon on the road. It was propelled by pulling two handles towards the rider in the manner of oars in a boat.

By 1887 the firm included in its catalogue the ordinary or 'high' bicycle, the improved Bicyclette safety cycle, the Rotary Tricycle and a quadricycle, all of which were selling well. It was this fact that enabled Lawson, who was then still working for the firm, to persuade Woodcock that the time was ripe for a further increase in the company's shareholdings. Thus, on August 4, the concern was transformed into a public company and the name changed to the Rudge Cycle Co Ltd, its first public prospectus appearing in the October of that year. The capital of the newly created company was a healthy £200,000, a sum of money that would have been beyond the wildest dreams of the old landlord of the 'Tiger's Head' at Wolverhampton, whose name the company bore.

In 1889 the firm attained one of the pinnacles of success of the day, a Gold Medal at the International Exhibition. By 1890 though, things were not looking quite so rosy in the cycle trade. The once lucrative American cycle-export market was being rapidly eroded by excellent locally built products having 'as-standard' items in their specifications which would be optional extras on British machines, but at lower overall prices. By the end of that year, the British cycle manufacturing industry was in the throes of an economic slump, whose resultant difficulties were to be made worse, for the Rudge concern, by the death on May 18 1891 of George Woodcock, its leading light. He had for long been driving himself too hard, almost without knowing it. He, like Dan Rudge before him, also died on licensed premises, at the 'King's Head' Hotel, Coventry, which he had owned for a number of years, and at the comparatively early age of 54, bequeathing a large personal fortune to his wife.

So ended the amazing and phenomenally successful career of England's first mass-producer of bicycles. He undoubtedly introduced a new concept into the industry. Under his direction, standardisation became the watchword and was insisted upon as never before. His approach certainly worked for, at his death, despite the recession in trade, he left the Rudge concern one of the most stable and prosperous cycle firms in the country.

But 1891 was a significant year in Rudge history from another viewpoint because it had also seen the founding of yet another cycle marque, one which was to have a profound influence on the firm's subsequent development. This was the Whitworth, the product of the Whitworth Cycle Company set up on June 1891 by Messrs Pugh & Sons, an old, established firm involved in the manufacture of screws, hinges and general builders' ironwork. It was owned by Charles Henry Pugh and had works in Rea Street, Digbeth, Birmingham.

The trademark chosen by Pugh for his new cycle

Charles Henry Pugh, who in June 1891 founded The Whitworth Cycle Company — one of the precursors of Rudge Whitworth Ltd.

company, and officially registered as such in 1891, depicted a red hand superimposed on a cycle wheel. The Works Manager of the new firm was Charles Henry Pugh's nephew, a young and lively engineer named Charles Vernon Pugh. Using new plant and the latest production methods, the firm proved an immediate success and within six months was ripe for expansion. Larger works were therefore obtained, again in Rea Street, to accommodate the increased demand for the firm's products.

The same year, young Harry Rudge, inheriting his father Dan's interest in bicycles, went into business with Charles Arthur Wedge, to produce Rudge-Wedge bicycles.

Whilst all this was going on, the Rudge concern had been having a difficult time. One of the most progressive engineers in the company at this time was Bob Walker Smith who, as assistant to 'Black' Draper (called 'Black' because of his dark complexion), the Works' Manager, had gradually established quite a name for himself as a designer. He had served his engineering apprenticeship back in the 1870s at the Railway Workshops in Wolverhampton. Later he had joined Dan Rudge's cycle concern in that town and had moved to Coventry with the other Rudge employees when George Woodcock bought the firm from Dan Rudge's widow in 1880.

Smith had designed the first Rudge tandem in 1886, the Olympian tandem for Marriott & Cooper in 1887, which was produced for M & C by Rudge, and also many of the cycle accessories that the Rudge concern had produced for Perry & Company during the 1880s. Albert Eadie was the Manager of the Perry Cycle Department, so the two got to know each other quite well over the years.

In 1892, Smith left the Rudge concern in Coventry and set up in business with Albert Eadie. They had acquired the long-established concern of George Townsend & Company at Hunt End near Redditch and laid the foundations of what was to become the Enfield Cycle Company. Then, in 1893, Rudge's General Manager, Walter Phillips, left to join the Humber concern which was a direct competitor. Thus within the space of two years the firm lost three key men. It was in a bad way and was looking for a lifeline. That lifeline was to be provided by the new and vigorous Whitworth Cycle Company which, in 1893, had been turned into a private limited company, under that name, to allow for further expansion.

Charles Vernon Pugh initiated a series of negotiations with the ailing Rudge company with the outcome that in October 1894 the two concerns merged under the name of Rudge-Whitworth Ltd, with Charles Vernon Pugh as Managing Director and J. Wallis (a former Rudge company man) as Chairman. Headquarters were to be at Crow Lane, Coventry, but Charles H. Pugh Ltd was to continue as a separate concern in Birmingham, manufacturing frames and other parts for Rudge-Whitworth in addition to its own items of ironmongery. However, in 1893, the old Rudge company had set up a drawing office and its sales organisation was clearly superior to that of the former Whitworth concern, so in December 1895 the office and staff of the latter were moved to Coventry as well. Amongst them was the new firm's Works Manager, John Vernon Pugh, a bachelor, who moved into a small flat over the Rudge-Whitworth works in Spon Street, Coventry.

Like most cycle companies of this period, Rudge-Whitworth was not averse to a little 'badge' engineering and in 1894 marketed a cut-price bicycle called the 'Crescent' at the very low price of £6 15s. It was felt undesirable to associate the Rudge name with such a low-price machine.

Despite his title of Works Manager, John Pugh had little to do with the day-to-day running of the factory, this being looked after by Victor Holroyd, the Works Superintendent, who had formerly been with Cycle Components of Birmingham.

After having left the former Rudge concern, Harry Lawson, in concert with a certain Terrah Hooley, got involved in a number of highly dubious cycle company flotations that would almost certainly have resulted in his being prosecuted under the law and convicted if they had happened today. The pair had floated a large number of companies whose profits could not be sustained, over-capitalising them and making profits out of the gullible public. This further enhanced the already bad slump in the cycle trade. Then, to cap everything, the American cycle manufacturers, having satisfied their home market, started to export their mass-produced machines to Britain and sell them at around two-thirds the price of English bicycles. The Rudge-Whitworth Board of Directors decided to counter this by establishing, in 1898, a small range of standard machines and almost halving their retail price. Other cycle makers soon followed suit and this effectively squeezed out their American rivals. So the situation was saved. At that time the Rudge-Whitworth concern was producing bicycles at the rate of some 27,000 a year. Also in 1898, the company introduced another innovation, a form of hire-purchase known as the 'Easy Payments Scheme' involving an initial down-payment of a guinea (£1.05) with the rest of the money being paid over a period of 11 months in equal monthly instalments.

By 1900 the demand for Rudge-Whitworth bicycles had risen so dramatically that the capacity of the existing plant was proving inadequate and the works in Crow Lane was extended. The new building linked the old 'Ariel' cycle factory with the more modern two-storeyed building in Trafalgar Street. Unfortunately on December 31 that year, the River Sherborne, a small stream that ran alongside the works, rose after heavy rain, came up over its banks and flooded the surrounding area. The Despatch Department of the Rudge Works was soon awash and many thousands of pounds worth of damage resulted. John Pugh decided to carry out a personal inspection and nearly had a serious accident in the process when he inadvertently stepped on to an unsupported piece of the planking laid down to enable workers to walk through that department without getting wet. He ended up in the flooded 10 ft (3 m) deep lift well, but fortunately suffered no serious injury. Enough was enough, he decided, the Despatch Department would from now on have to go up to the first floor to avoid the recurrence of such inconvenience. A happier event that year for John Pugh was his appointment by the Board to the post of Works' Director for his highly successful efforts in expanding the company. Victor Holroyd stepped up to become Works' Manager — whose functions he performed anyway.

Quite a few cycle concerns at this time were becoming interested in motorising their products or in getting involved generally in the new motorcycle market. Now the Werner Frère Company of Paris had gained a good reputation for itself in the fledgling motorcycle industry, so Charles Vernon Pugh negotiated an agreement with that firm's UK subsidiary, Werner Motors Ltd, for Rudge-Whitworth to become the sole agent for Werner motorcycles in South Africa. Unfortunately, it was soon discovered that the South African Rudge-Whitworth sales depot, located in Johannesburg, was having the prices of its newly imported Werner motorcycles undercut to the extent of £5 per machine by other South African importers selling new Werner machines in that country. Clearly this situation was intolerable for Charles Pugh. He refused to pay the balance owing to Werner Motors Ltd and that concern responded by taking him to court over the matter. The case, which was heard in the King's Bench Division before the Lord Chief Justice, ended with the latter recommending a postponement of the case while the two parties met to discuss the affair. The upshot was that they settled out of court, but this fiasco tended somewhat to prejudice Charles Pugh and the Rudge-Whitworth concern against any further immediate involvement in the motorcycle market.

By 1902, Rudge-Whitworth had established a Chemical and Physical Laboratory, with H. L. Heathcote in charge, to conduct research into new materials and to test existing cycle parts. Meanwhile, in Wolverhampton, the Rudge-Wedge concern had prospered and in 1903 decided to produce its first motorcycle, the first prototype of which appeared on the stand of H.M. Hobson at the Crystal Palace Cycle Show in 1904. Powered by a Minerva engine clipped to the front down tube of the frame, the machine had a most unusual suspension system. The frame was in two parts, one carrying the saddle, handlebars and pedals, the other carrying the engine and wheels. There was a common steering head, the front forks being rigid while the two sub-frames were attached to each other at four points by springs. Not a happy arrangement, since the springs were undamped and very rapidly induced nausea in the rider when travelling over rough road surfaces.

In 1904 the cycle industry entered another depressed phase and the Rudge-Wedge concern found itself in financial difficulties. Rudge-Whitworth bought the company out and the Wedge name lived on for a while as the name for one of the Rudge-Whitworth cycles of that period. Rudge-Whitworth were not yet, however, inclined to re-enter the motorcycle market. The following year (1905) saw the company's output of machines exceed 50,000 for the first time — more than the combined output of the previous two years. During the latter half of 1905 and the early part of 1906, additional factory buildings were constructed in Spon Street, Coventry, to accommodate this rapidly increasing production. By the end of 1906 the production figures had soared to 75,000 machines a year and were still rising rapidly. At the same time the range of machines had been extended to give 75 different models, each with a ten-year guarantee.

The cycle trade and certain cycling journals, notably *The Cycle Trader*, were becoming extremely critical of the Rudge-Whitworth depot system at this time. It was considered a form of direct trading liable to take business away from the local cycle traders. As a result of the remarks by the aforementioned journal, early in 1905 the company obtained a writ against it. The magazine, however, apologised and withdrew all statements it had made on the subject during the previous year and the matter did not actually come to court. In fact the rug had been pulled out from under its defence case, by Rudge-Whitworth Ltd becoming a member of the Cycle and Allied Trades Association, which had decided that no unfair trading was involved in the Rudge depot system.

On the sporting front, the Rudge concern was now receiving excellent publicity for its products. A wheel-building part of the business had developed which was now supplying proprietary detachable car wheels as well. Immediately prior to the official opening of the Brooklands racing track at Weybridge in Surrey, the previous year (1907), ex-Rudge man Selwyn F. Edge had taken his Napier racing car, shod with Rudge-Whitworth wheels, round the circuit on a gruelling record-breaking run to become the first car driver to achieve an average of 60 mph (96.6 km/h) for 24 hours. The resultant publicity produced an upsurge in the popularity of lightweight Rudge-Whitworth wheels amongst car-racing men. Thus in the 1908 Car TT Race in the Isle of Man, 21 of the 35 starters were using them.

The year 1908 also saw the launching of a company house journal, *The Rudge Record*, designed to keep all Rudge employees in touch with what was going on in the company. This was quite an unusual happening for a company of that period, but Rudge-Whitworth had its own printing press for turning out catalogues so it was a natural progression to publish an 'in-house' journal as well. Produced by the Advertising Department with J.F. Tidswell as Editor, the journal was first suggested by a gentleman of the name of Skilton, in the Sales Department. It proved a great success. Amongst some of the happenings included in its earlier editions were the record-breaking activities of the Rudge-Whitworth Cycle Club members and various social functions including the firm's annual garden party for employees held every June by the Chairman, Charles Vernon Pugh, and his wife, at their home at Radford.

The year 1908 ended well for the company with a declared profit, at the Annual General Meeting in November, of £15,860. This was less than the previous year, but the firm's assets had now risen to a healthy £401,506. Shareholders were well pleased. At the same meeting it was also announced that the Regent Street depot had been closed and a new one had been opened at 230 Tottenham Court Road, London, only a stone's throw away from Warren Street, then regarded as being the Mecca of motoring in London.

2

Power and pedal (1909-1911)

It had become obvious to the Board of Directors of Rudge-Whitworth that it would soon have to reconsider its position concerning motorised transport. By 1909 there were quite a few motor cars about on the roads of Britain, even if they were, by and large, beyond the financial means of the ordinary person. However, it did not take too much imagination to see that motorcycles would soon be making inroads into the bicycle market.

When other cycle makers had started producing powered vehicles at the beginning of the century, many had plumped for the three-wheeler because of the handling problems associated with motorised bicycles. There were then only a few proprietary engines available and these, when fitted into cycles, produced machines of uncertain and sometimes dangerous handling characteristics. Furthermore, cycle makers were unwilling to invest in the expensive manufacture of their own engines and to design their own purpose-built machines, as the market seemed so uncertain. In contrast, by 1909 much of the ground work on motorcycle frame design had been done, the principles established and the market prospects for motorcycles were much firmer. It was because of this that in 1909 the Rudge-Whitworth Board was persuaded to initiate development work on a prototype Rudge-Whitworth motorcycle for the coming year.

The first problem was to decide on a suitable engine configuration. At that time the vast majority of machines had side-valve engines, but the few overhead-valve engines that were produced (notably by the J.A. Prestwich concern of Tottenham) offered greater power potential for their size due to a more compact combustion chamber. This aspect had to be borne in mind since motorcyclists were very interested in performance, as well as reliability, and motorcycle racing at Brooklands and in the Isle of Man was now taking off in a big way. If the Rudge-Whitworth motorcycle was to obtain good publicity in racing it would have to have an engine that could compete successfully with its rivals on a power basis.

The Achilles' heel of early motorcycle engines was the unreliability of their valves, in particular their exhaust valves. If these broke under continuous heavy-load running conditions, as they often did in the early days with the slotted valve stems and inferior valve materials used, the result could be a seriously damaged engine when using the overhead-valve (ohv) configuration. With a side-valve (sv), on the other hand, all that would happen under such circumstances was that the engine would cease to function. This presented something of a dilemma to John Pugh for he wanted performance and reliability. His answer was to strike a compromise by having a side exhaust valve and housing the inlet valve immediately above it where, if the valve head broke off, it could do the least damage. This eliminated the valve-breakage hazards associated with the ohv engine, but at the same time produced a more efficient engine than the side-valve design. There were no satisfactory proprietary engines of this type so the engine would have to be produced entirely by the Rudge-Whitworth factory.

During 1909, much secret experimental work on the project was carried out. Other makes of motorcycle were examined and a stream of new patents was granted to the Rudge-Whitworth concern, in many cases based on completely new answers to then existing motorcycle design problems. Two features patented at this time were to have a very long production life. They were the one-piece front fork shackle (British Patent 546796/1909) and the famous Rudge enclosed front-fork spring assembly, which were both used on the firm's machines right up until motorcycle production for the public ceased in 1939. A patented, hinged, rear mudguard flap was another innovation, intended to facilitate wheel removal in the event of a rear tyre puncture — an all too common occurrence on anything other than quite short motorcycle journeys in those days, due to the poor state of the roads and frequent presence in them of sharp stones, nails and sharp-edged horseshoes.

By 1909 the '3½ hp' single-cylinder machine had emerged as being by far the most popular size of motorcycle with the buying public. This was for a variety of reasons. The nominal horsepower, '3½', corresponded to an engine capacity of 500cc. This was at that time considered to be the largest practicable cylinder size for a road-going machine and a single-cylinder machine of that capacity was considerably easier to maintain, much lighter and less expensive to buy than a twin-cylinder. For these reasons, the new Rudge-Whitworth was given an inlet-over-exhaust (ioe) valve, single-cylinder engine, having a bore and stroke of 85 × 88 mm, giving it a swept volume of 499cc. A number of changes of detail design occurred before the machine eventually went into production.

An initial assessment of the design, clearly based on verbal information, appeared in *The Motor Cycle* of June 30 1910. In contrast to the customary engineering practice of the day, a separate cast-iron cylinder head and cylinder barrel were used, together with a cast-iron piston. According to *The Motor Cycle* report: ' . . . The detachable head is secured to the cylinder by four long bolts, let into the crankcase in a similar manner to the old De Dion and MMC tricycle engines . . . '

On the first prototype motorcycle, however, which unbeknown to *The Motor Cycle* was assembled on July 27 1910 from blueprints completed only nine days before, there were only three holding-down bolts used. This was because the prototype engine had its exhaust valve chest located at the front offside of the cylinder, to facilitate exhaust-valve cooling. Such an arrangement made a four-bolt fixing impracticable.

The exhaust valve and its guide were located in a plate bolted to the top of the exhaust valve chest. The inlet valve, as described in the original report in the motorcycling press, was held down on its seating by a dome provided with a bayonet joint. This idea had obviously proved unsatisfactory, since the prototype machine employed a central retaining nut through which passed a short pushrod from the overhead rocker.

An unconventional feature of the engine design was the method of valve actuation employed. A long pushrod passing up from the front-mounted cam box on the offside of the magneto (mounted on a platform in front of the engine) actuated the overhead inlet valve; it was adjustable for length to provide tappet clearance adjustment. On the driven spindle of the magneto was mounted a single cam which served to operate, via followers, both the inlet and exhaust valve. The offside chain-drive to the half-engine-speed front-mounted magneto was fully enclosed. Within this magneto drive cover was located an exhaust-valve lifter working on the cam follower. This provided compression release to assist starting the machine.

A special Brown & Barlow spray-type carburettor, made especially to a Rudge-Whitworth design, was mounted behind the engine and fed the inlet valve via a long horizontal induction pipe passing between the lower tank rail and the top of the cylinder head. To enable a low seating position to be achieved, the top frame tube ran parallel to the lower one for only two-thirds of its length. It then curved downwards towards the back stays and seat tube lug. The resultant distance from the top of the saddle to the ground was a mere 30.75 in (780 mm).

In other respects the new prototype Rudge-Whitworth was on standard motorcycle lines for the day in having: a V-pulley on the engine shaft connected by a belt direct to the rear-wheel pulley; pedalling gear; and braking by the usual belt-rim applied back brake plus a hand-applied rim type on the front wheel. There was a luggage carrier over the rear mudguard consisting of a strengthened steel plate supported by the customary tubular framework and stays.

The tank was provided with a combined petrol gauge and filter, and the filler caps for the front 0.5-gallon (2.27-litre) oil and rear 1.35-gallon (6.61-litre) petrol compartments were large enough to enable ordinary car funnels to be used. It followed the usual slim lines of the period. There were two petrol taps, one supplying the carburettor and the other the priming tap for aiding cold starting.

A vertical tube passed through the tank immediately above the inlet valve, enabling both inlet and exhaust valves to be ground in 'in-position'. This is not nowadays regarded as being a desirable practice, since grinding paste and metal powder left over from the grinding operation are difficult to remove and could produce increased engine wear.

A further prototype (registered as DU 4219) was completed on August 19 and another, DU 4916, eight days later, each machine weighing 208 lb (99 kg). At the same time, the Rudge-Whitworth design department had evolved a novel combined multi-

Section of the engine shaft multiple-plate clutch used on the 1910-11 Rudge-Whitworth motorcycles.

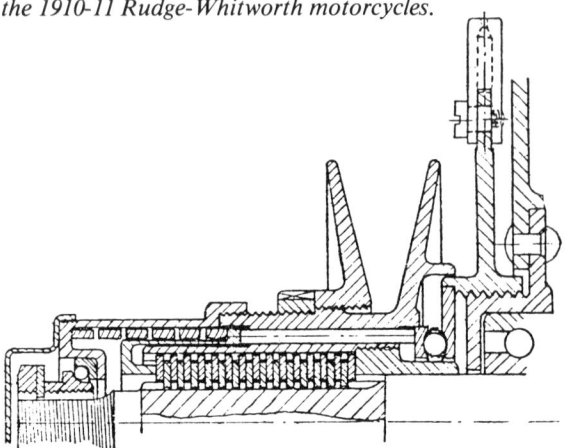

A close-up of the engine of Charles Burney's Brookands' Rudge, which was fitted with one of the new variable-jet Brown & Barlow carburettors and had an exceptionally long air intake.

plate clutch and adjustable pulley design, which was granted British Patent 26634/1910.

One of the problems with the first engine design was that it was difficult to achieve efficient gas sealing between the cylinder head and barrel with only three points of fixing. This led to excessive bolt tightening which caused distortion. On September 1 1910, a new cylinder head design was produced, involving strengthening ribs between the retaining-bolt holes to resist distortion. This solved the problem but involved moving the priming tap so far from the sparking plug, which was located near the exhaust valve, as to make it ineffective as an aid to starting.

At the end of September a third redesign occurred, the inlet valve mounting and cylinder head this time being made in one casting. The inlet valve and induction pipe were now the only detachable portions of the head. The priming tap was more central in this design, so starting was unimpaired, and the inlet rocker now bore directly upon the valve. A big change was the adoption of a train of gears to drive the magneto, each valve being operated by its own cam. There had also been a considerable tidying up of the design. Thus the oil drain plug had now been moved from its former vulnerable and awkward location at the bottom of the crankcase to a safer position on its offside.

On October 18 1910, a lightweight stripped version of this design was produced, weighing a mere 104 lb (41 kg). It was entered in the Auto Cycle Union's Round Midlands Tour and performed reliably if modestly. On Friday, November 4, F. Wright, Charles S. Burney and the young 21-year-old rider, Victor Surridge, son of the famous Warwickshire cricketer of the day, took three of the new 'third design' Rudge-Whitworths down to Brooklands Track, at Weybridge in Surrey, and carried out extensive testing on them. The three machines were subsequently displayed at the Olympia Cycle & Motorcycle Show towards the end of the year in November. The track tests, however, showed that the performance of this third design was inadequate, so work was put in hand on a fourth design with better performance in mind.

When the first Rudge engine was designed, it was realised that there had to be no possibility of any broken-off valve head entering the cylinder. Then, as designed, the passage linking the valve chamber and the cylinder was of rectangular cross-section measuring 45.7×20.3 mm giving an area of 9.28 cm^2, while the inlet valve's head diameter was 44.5 mm. Subsequently, by altering the shape but keeping the same large cross-sectional area, the maximum width of the passage was reduced to 43 mm, thereby preventing the possibility of a broken valve head entering the cylinder. This obsession with valve head problems had deflected attention from the need for good engine breathing. This problem was solved by employing a larger inlet port and this modification was carried out on the fourth pre-production design, drawings of which were completed by November 2 1910.

On this fourth prototype there was a completely new design of connecting rod and crankpin, enabling the fitting of roller-bearing big and small ends instead of plain-bearing types as on the three previous designs. This change was intended to help cut down the frictional losses in the engine and enable greater power to be developed at the crankshaft. At the same time the exhaust-valve lifter was moved from inside the timing case to operate on the exhaust valve itself and was attached to one of the cylinder barrel holding-down bolts. The exhaust port outlet had been enlarged to enable the exhaust pipe to be located inside it instead of using the threaded-stub and gland-nut method adopted on the previous design and which was prone to distortion, making exhaust pipe removal

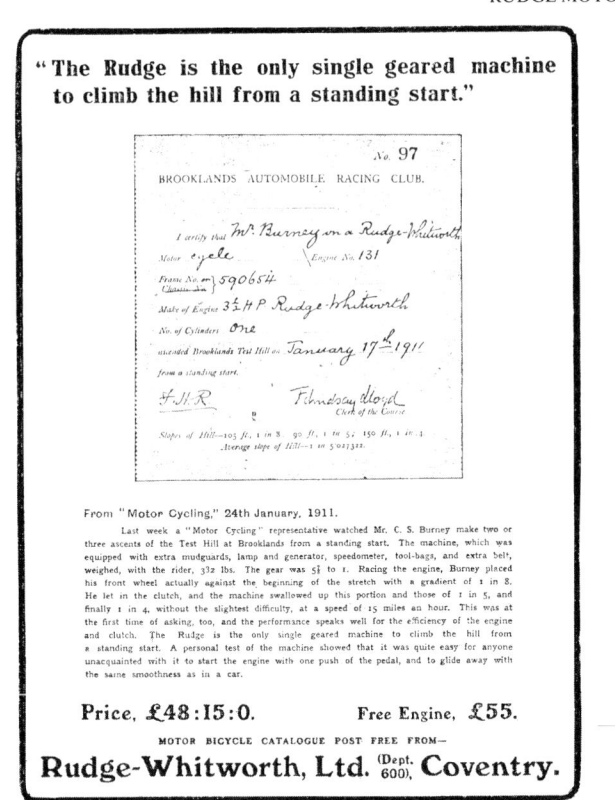

In January 1911, Charles Burney made a successful ascent of the Brooklands Test Hill on his 499cc Rudge.

A Rudge-Whitworth advertisement that appeared in The Motor Cycle *magazine in January 1911.*

difficult after a time. The fourth design also featured a horizontal magneto platform and the overhead rocker pillar had been strengthened. A new timing-side crankcase half had been necessitated by some of these changes, so the opportunity was taken to include the word 'Rudge' on the casting just below the timing cover, a feature that was to be retained until 1915.

The first, completed, fourth-design machine was ready for testing on November 20 1910. Six more were made the following month, including three with clutches for use by works' riders as competition machines and weighing 182 lb (82.7 kg) apiece.

The new machines occupied prominent positions on Stands 54 and 55 at the Olympia Motorcycle Show during the week of November 21 to 26 that year. Meanwhile large double-page advertisements appeared in the motorcycling press describing the proposed new production Rudge-Whitworth models of which there were two — first, a direct drive model priced at £48 15s and second, a free-engine sporting model incorporating the new patented clutch and costing £55. The fixed gear on both models could be altered from 4:1 down to 6:1 by means of an adjustable pulley. A 1½-gallon (6.8-litre) petrol tank was now fitted having an internal dam to regulate the petrol supply. Once again there was a drip feed to the compression tap to assist starting. An interesting feature was the adoption of a spring-operated stand for the rear wheel. As with the cycle range and the experimental motorcycles, the new production Rudge-Whitworth machines were to have an attractive black cellulose finish on the handlebars, a characteristic Rudge feature over the years. Manufacture of these motorcycles for the general public commenced on January 11 1911.

John Pugh, realising the value of sporting successes on the advertising front, persuaded his Board of Directors to allow a few of the prototype machines to be tried out in sporting and other activities right from the start. Thus on January 17, Charles Burney brought his Mark IV development model (built on December 12 1910) down to the track and successfully took it up the formidable Brooklands Test Hill with its 1-in-5 mean gradient, clearing the top at a steady 15 mph (24 km/h).

Early sporting activity for the new Rudge-

> **The Run-away Rudge**
>
> At Brooklands, 18th March.
>
> In the First Senior One Hour Tourist Trophy Race the Rudge 3½ h.p. T.T. beat every other single-cylinder machine by nearly a mile.
>
> | RUDGE | 55 miles 496 yds. in the hour. |
> |---|---|
> | Nearest other make | 54 miles 587 yds. in the hour. |
>
> MOTOR CYCLING, March 21st, says: "The remarkable way in which the Rudge was leading the other single-cylinders was commented on by everyone."
>
> PRICE.
> Roadster or T.T. Model | Fixed Engine £48 15 0 — Free Engine Model - £55 0 0 (With Multi-plate Clutch and Pedal Engine Starter).
>
> THE FULL CATALOGUE NOW READY—POST FREE FROM
> **RUDGE-WHITWORTH, LTD.,** Dept. 600, COVENTRY.

Rudge advertisement celebrating Victor Surridge's achievement in the Brooklands' One-Hour TT Race, 1911. He finished second, having covered 55 miles 496 yards.

Whitworth motorcycles was not just confined to speed tests, but also road and reliability trials. An early Rudge protagonist here was the private owner B. Alan Hill, who was lucky enough to be allowed a prototype model to ride in the Motor Cycling Club's London-to-Exeter Winter Run in which he produced a creditable performance. In January 1911 he produced a similar performance in the ACU's Quarterly Trial held over a course from Croydon to Hastings and back. Three Rudge-Whitworths were entered, piloted by Hill, Charles Burney and Victor Surridge. All three made non-stop runs. Burney's engine was fitted with a special double-sparking-plug arrangement controlled by a two-way switch, so that either plug could be brought into action without dismounting. Another modification, this time on all three machines, was the adoption of a neat spring clip to retain the exhaust pipe.

An ingenious form of pedal starting for the engine was introduced towards the end of January on the free-engine models. It consisted of a small conical friction clutch on the camshaft, worked by the ordinary chain and pedal gear. The clutch itself consisted of a fluted, hardened cone on the end of the camshaft which engaged with a corresponding tapered hole in the centre of a hardened bush. This bush was thrown in and out of engagement with the camshaft by the lateral movement of a two-start square thread, which forced the two tapered surfaces together with sufficient contact friction to start the engine on the forward stroke of the pedals. On the reverse stroke the square thread disengaged the clutch by unscrewing the bush. Charles Burney's Quarterly Trials machine was fitted with the device and also an aluminium guard to prevent the rider's foot fouling the clutch mechanism on the nearside of the machine.

Quite early on, the advantage of being able to lift not only the back wheel, but also the front wheel, off the ground to repair a puncture, rather than resting the vulnerable bottom of the crankcase on a brick, piece of wood or empty petrol can, became apparent. As a result, a front wheel stand was also added to the specification of Rudge-Whitworth machines on January 23 1911.

The much improved performance of the new sports machines encouraged the production of a TT-type model and, on March 1 1911, two machines of this type were built despite the Rudge-Whitworth concern having never entered the Isle of Man Races before. These were immediately sent off to the firm's Dublin Depot for testing on local Irish roads where the constabulary were more tolerant of motorcycle speedsters than in England. As a result A. Carvill, riding one of them, gained a fastest time of the day award in the Dublin & District Motor Cycle Club's Hill Climb at Ballinslaughter on Friday, March 17. Ten days after this sporting success, Joseph Healey started on an officially observed, six-day, non-stop endurance run from Dublin to Belfast and back, covering a total distance of 1296 miles (2087 km) without a single breakdown on his Rudge.

The next appearance of Victor Surridge and Charles Burney at Brooklands was at the First BMCRC Race Meeting of 1911 held on Saturday, March 18. Burney's machine had been converted to rigid forks for the occasion as was the custom with some track-racing men of the day. Surridge's machine, on the other hand, retained the standard pattern Rudge-Whitworth sprung front forks and showed, not surprisingly, much better handling. He managed to take his machine home into second place behind Sidney Tessier's 580cc BAT-JAP twin, in the Senior One-Hour Brooklands TT Race, covering 55 miles 496 yards (89.02 km). A fine performance for a virtually standard '500' in those days.

60 Miles in the HOUR

THE
3½ h.p. RUDGE
IS THE
First and only machine
to beat 60 miles in the hour.
WORLD'S RECORD.

Accomplished by V. J. Surridge,
at Brooklands, on May 25/11.

A new edition of the Motor Bicycle Catalogue, free by post, fully describes the many exclusive features of the Rudge, and should be read by every actual or prospective motor cyclist.

PRICES.
Fixed Engine, £48 15/-   T.T. Model, £48 15/-
Free Engine (with multi-plate clutch and pedal starting gear), £55.

RUDGE-WHITWORTH, Ltd. (Dept. 600), COVENTRY.
London Depots where demonstrations are arranged—
230, Tottenham Court Road, W.   23, Holborn Viaduct, E.C.

Surridge's successful hour record provided superb publicity for the new Rudge-Whitworth machines.

It was this performance, as much as anything, that persuaded John Pugh to allow Victor Surridge and Charles Burney to have a crack on their Rudges at the 500cc one-hour record. Up to that time no machine of under half a litre engine capacity had successfully completed, non-stop, 60 miles in the hour. It was not for want of trying either and there were a number of contenders for the title of being the first to do so, including Billy Newsome on the works Triumph and the formidable George Stanley on the works Singer.

The first opportunity for the Rudge duo to try for the record arose in the Senior One-Hour Brooklands TT Race at the BMCRC's second monthly meeting on Wednesday, April 26 1911. Unfortunately Billy Newsome's Triumph proved the fastest of the 500 singles and in finishing second to Charlie Collier's 580cc twin Matchless he succeeded in breaking the record with a distance of 59 miles 1478 yards (96.36 km). Charles Burney was third on his Rudge, which had covered 57 miles 869 yards (92.58 km).

Billy Newsome did not have things all his own way though and Victor Surridge made a strong fight for second place. He lost too much time, however, after a stop for petrol, to get back amongst the leaders before the end of the race hove in sight. Nevertheless, he managed to average a useful 62.71 mph (100.98 km/h) for all of his flying laps prior to his retiring from the race on his 15th lap.

Clearly both Victor and Charles were within striking distance of the record, but if a Rudge-Whitworth was to become the first 500cc machine to achieve the magic 'mile-a-minute' average for the hour, a successful record bid would have to be made soon. Meanwhile all racing opportunities at Brooklands that presented good advertising potential, were taken advantage of. The next of these was a mid-week BARC car race meeting, held on May 10 1911, at which there were two motorcycle events. In the first of these, a Short Motor Cycle Handicap over two laps, Victor Surridge gained third place starting from the 58-seconds-from-scratch position.

On the Friday of the same week (May 12), Victor set out in earnest at the track on the trail of 500cc solo records. It was an ideal Spring day for record breaking, with the sun shining and a gentle breeze blowing from the north-west.

His first attempt was on the flying-start kilometre record held by Harold Bowen's BAT-JAP. Victor's Rudge was in excellent fettle and he went through the measured strip on the Railway Straight to set a new world record at 65.79 mph (105.94 km/h). Unfortunately, though, through a misunderstanding with the timekeepers, he also made a second and unnecessary attempt on the kilometre (short-distance records then being one-way affairs), easing up before reaching the mile-post during what was ostensibly a flying-start mile record attempt. Realising his error, he decided to have another attempt at the mile distance but, unwisely, returned the wrong way round the track against the normal direction of racing to do so, and collided with a car which suddenly appeared from behind one of the Brooklands aeroplane sheds. Fortunately he was not injured, but he had badly damaged the front forks of his machine. This resulted in an hour's delay in the proceedings while they were being replaced. This being done, he promptly went out and beat John Gibson's mile record on the Triumph by some 2 mph (3.2 km/h) with a speed of 66.18 mph (106.57 km/h).

The time was now ripe for him to attack the hour record, and, armed with a spare Lyso driving belt and spare petrol pipe, both wired to his front forks, he started out on his attempt at 5.50 pm. After two laps a broken exhaust valve forced a halt, then at 6.20 pm with a new valve in place, he made a fresh

Above *Here is Victor Surridge waiting for Mr Whitney Straight of the ACU to give him the 'go ahead' prior to his successful attempt at becoming the first to cover 60 miles in the hour on a '500'.*

Below *Victor Surridge on Friday, May 12 1911, after breaking the 500cc solo flying-kilometre record at over 66 mph.*

start on his attempt. With 19 laps to his credit everything looked set for a successful conclusion to the day's proceedings, but then, on his 20th circuit with only 6 min to go before his time was up, fate struck again. An inlet cotter broke, the valve spring disappearing somewhere out on the track never to be seen again and that was that. Victor would have to await another day to see if he would achieve the elusive '60 in the hour'. There was a consolation though, in that he had set up a new record for 50 miles at an average speed of 58.83 mph (94.73 km/h).

It is interesting that the engine of this machine was said to have been churning out some 8.5 bhp at 2000 rpm at this stage of its development.

Clearly world records, as many a Brooklands veteran could have told Victor, did not come easily, so both he and the Rudge team newcomer, Billy Elce, decided to do a little more racing at the track before making any further attempts on the hour. They decided to enter their machines in the BMCRC's Third 1911 Monthly Race Meeting held the following Wednesday (May 17). The day's events again provided Rudge-Whitworth with useful publicity and a good opportunity for carrying out further engine tuning for the hour-record attempt. During the course of the Record Time Trials, held over the Railway Straight, Victor

Billy Elce (pronounced 'Elsie') was another highly successful Brooklands Rudge exponent in the years prior to the 1914-18 war.

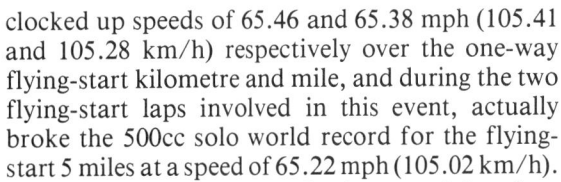

Japanese rider K. Yano (BAT) and W. Stanhope Spencer (Rudge) on the starting line at the Brooklands' 1911 May Race Meeting of the BMCRC.

clocked up speeds of 65.46 and 65.38 mph (105.41 and 105.28 km/h) respectively over the one-way flying-start kilometre and mile, and during the two flying-start laps involved in this event, actually broke the 500cc solo world record for the flying-start 5 miles at a speed of 65.22 mph (105.02 km/h).

It was, however, in the final event of the afternoon, the three-lap 500cc Solo Scratch Race for the Palmer Tyre Company's Silver Cup, that he really shone. At the end of the first lap he was running neck-and-neck with Billy Newsome on the works Triumph, having set a new Brooklands 500cc solo standing-start lap-record of 60.62 mph (97.62 km/h). It was a close tussle, but by the end of the second lap Surridge had a slight lead. On the last lap the two men had a titanic struggle for the lead, first one and then the other holding it, until just before the end Victor got his wheel ahead to win by a mere 2.4 seconds — his average speed was a healthy 63.60 mph (102.42 km/h).

So close had Victor Surridge been to breaking the 500cc hour record earlier in the month that it seemed as if he could just not fail to take it if he persisted — clearly the machine was fast enough. He made the necessary arrangements with Major Lindsay Lloyd, the Brooklands Clerk of the Course, and on Thursday, May 25 1911, sallied forth once again, this time much earlier in the day, at 3 pm.

It was a brilliantly sunny day but, unfortunately, the resultant glare of reflected sunlight thrown up from the track, experienced during a warming-up lap, forced him to postpone matters for an hour until it had abated. Then, when conditions had considerably improved, he got under way.

Foolishly, on such an important record attempt, an experimental exhaust valve had been fitted on his engine. This broke after only five laps and forced a halt. With a new valve fitted Victor started on yet another attempt at the record, but with a standing-start lap at only 55 mph (88.57 km/h) and his second at only 58.5 mph (94.20 km/h), the pace was much too slow. Wisely he stopped for engine adjustments and the whole attempt was restarted.

Things now began well and Victor tried to maintain a record-breaking average, but found it difficult to be consistent in his lap times. On his 20th lap he stopped for oil and made up for this by going like the wind on his next lap and taking the Brooklands 500cc solo flying-lap record in the process at 66.47 mph (107.04 km/h). Covering his last two laps at 65 and 66 mph (104.7 and 106.3 km/h), he completed 60 miles 783 yards (97.33 km) in the hour to become the first rider in history to achieve 60 miles in that time on a 500cc solo. Apart from an extra-long fuel tank specially tailored for the run, the machine was (externally at least) a stripped-down, standard Rudge-Whitworth motorcycle, with fixed, single-geared direct drive by Lyso belt to the rear wheel, fitted with a Brown & Barlow carburettor, Dunlop tyres and CAV magneto ignition.

To celebrate Victor Surridge's success, two weeks later on the evening of Thursday, June 8, the Rudge-Whitworth company gave a dinner party in his honour at the Savoy Hotel in London.

Meanwhile further honours had been gained by Rudge riders in reliability trials. Thus, in the London-to-Land's End Run, staged over the four days from April 14 to 17 1911 inclusive, G.T. Gray, B. Alan Hill, Billy Elce and A.J. Sproston all gained

Gold Medals, while Charles Burney won a silver. Two days later the reliability of Rudge machines came to the fore in the Spring Quarterly Trial at Sutton Bank in Yorkshire, with Hill finishing first in the heavyweight class against tough opposition.

For the 1911 Coronation Jubilee Tourist Trophy Races in the Isle of Man, coinciding with the accession of George V to the English throne and the 60th anniversary of the formation of the Isle of Man's Borough of Douglas, the Auto Cycle Union (ACU) had decided to run the event over a new, longer and much tougher circuit. The 'Short Course' used hitherto was getting much too easy to be a real test for the larger capacity machines, so the ACU's answer was to change to the 'Four Inch Course' used by the car racers. This used much of the present-day 'Mountain Circuit'. At the same time, the event was split up into separate races to be run on different days. The first of these, the Junior Race, was to be for single-cylinder machines up to 300cc and twins up to 340cc, and was to be staged on Friday, June 30, over four laps of the 37½-mile (60.3 km) circuit — a total of 150 miles (241.5 km). The Senior Race, for singles up to 500cc and multis up to 585cc, however, was to cover five laps totalling 187.5 miles (302 km). The Rudge-Whitworth Board decided that the firm should enter a team of four riders in the bigger class, in view of the firm's recent sporting successes and the kudos obtainable if they were to be successful in 'The Island'. The riders were Victor Surridge, John Gibson (a well-known Brooklands rider of the day), the Irish rider, Joseph Healy, and A.J. Sproston. In addition, there were five entries on Rudge machines by private owners, comprising H. Houghton, Vernon Taylor, J. Prendergast, Billy Elce and W. Stanhope Spencer.

The entire accommodation of the Glen Helen Hotel had been taken over by the Rudge-Whitworth concern, an outhouse being converted into a depot for its team. Close by the hotel, some rooms had been acquired and converted into a spare parts store and workshop for the duration of the racing fortnight. Thus the approach of the company to racing was quite business-like right from the start. Such an approach deserved success, but it was not to come in the Island that year, for during practice Victor Surridge had a fatal crash at Glen Helen, the

During the first five months of 1911, Rudge-Whitworth motorcycles gained a formidable list of sporting successes.

At the end of March 1911, Irishman Joseph Healy gave a clear demonstration of the reliability of the new 3½ hp Rudge motorcycle.

Only 21 years-old, Victor Surridge became the first Isle of Man TT motorcycle racing fatality during the practice period for the 1911 races.

first-ever TT fatality. Only four of the original nine Rudge entries, in fact, started the race and the first of these to finish was that ridden into 21st place by Stanhope Spencer, while John Gibson followed him home in 22nd spot, Prendergast and Sproston being forced to retire.

Meanwhile, back on the mainland, the factory was able to claim some sporting success with nine Rudges successfully completing the outward journey of the London-to-Edinburgh Run, while seven of the nine successfully completed the round trip. Amongst the Rudge riders was A. Mabon, whose patented variable gear was being fitted to a number of Rudge-Whitworth machines from the end of March 1911, in addition to the then popular German-made NSU two-speed gear.

The Mabon gear consisted of a chain-driven expandable pulley arranged upon a radial arm so that as the pulley expanded the arm was operated to bring the pulley further and further away from the belt rim to ensure uniform belt tension. The gear was controlled by a long lever which worked in a quadrant with five notches arranged under the tank on the nearside of the machine. One of the notches gave a 'free-engine' position, the other four notches yielding positive gears in proportionately decreasing ratios.

Of the Rudge-Whitworth machines entered in the 1911 Senior TT Race, some were single-geared, some were fitted with Armstrong-Triplex three-speed gears in the back hub, others had NSU two-speed gears mounted on the engine shaft, while at least one was fitted with a Mabon infinitely-variable gear. The most noticeable departure from standard on these machines was the use of large 'cellar-flap' type spring-closing filler caps for both the oil and petrol compartments of the tank, which carried 2 gallons (9.1 litres) of petrol and ½ gallon (2.3 litres) of oil.

The advantages of the new clutch, mentioned earlier, had become so apparent that, by May 1911, the fixed engine model had been dropped from production in its favour and could only be obtained by special order. The TT practising had amply demonstrated the advantages of some form of variable-speed gearing over the new and hilly 'Mountain' circuit. It was felt, however, that the existing variable-speed gear used by Rudge owners did not really fit the bill. The Armstrong-Triplex gear had only a very limited power handling capability as did also the NSU unit. The Mabon gear seemed to be on the right lines, but like the Zenith-Gradua gear before it, it did not solve the problem of how to achieve constant belt alignment despite providing more or less constant belt tension. This proved a restriction on the maximum possible range of ratios, increasing both power loss and belt wear.

After the Isle of Man Races, variable gear experiments continued. The event had clearly demonstrated the need for a variable gear with a wider range of ratios than the Mabon gear, which proved difficult to declutch under racing conditions and also wore rather rapidly. John Pugh was working on an answer to this problem at the time of the 1911 Motorcycle Show in November at Olympia — the subsequently famous 'Multi' gear. However, that did not come to fruition until the New Year. But this is to jump too far ahead in our story.

Earlier, on August 19, at the Nottingham Motor Club's Clipstone Speed Trials, young 21-year-old W. Stanhope Spencer had re-established the association of speed with the Rudge-Whitworth name by setting up a new 500cc solo flying-start mile world record at an average speed of 72.58 mph (116.79 km/h) and became the second rider of a '500' in history officially to exceed 70 mph (112.7 km/h), the first being Dan Bradbury earlier the same year (1911) on his 490cc Norton during the Sheffield and Hallamshire MCC Sprint Meeting.

Just before the Motorcycle Show each year, most of the popular motorcycle firms indulged in attempts at world speed records at Brooklands to round off the sporting year. Rudge-Whitworth was no exception to this rule and on Tuesday, October 3, Stanhope Spencer set out after 500cc-class records at the track, with Billy Elce and W.L.T. Rhys as his team mates. Billy soon retired with magneto trouble, while Rhys continued to lap at a little over 60 mph (97 km/h), which was too low a speed for successful record breaking. Spencer, however, was going well and settling down to a 66 mph (94 km/h) average. Despite breaking an inlet cotter pin shortly before the end of his scheduled two-hour run, he managed to set a number of new long-distance and period world records including the 500cc solo one-hour at 65.45 mph (105.39 km/h) and the two-hour at 61.60 mph (99.19 km/h).

After a break for lunch, Frank Pither set out to establish the first-ever 500cc sidecar record for the hour, driving a standard single-geared Rudge fitted with a Portland sidecar having a canoe-shaped wicker-work body. With an all-up weight of 548 lb (249 kg) this outfit kept up a steady 42 mph (68 km/h) average until a sparking plug blew out of the cylinder head, reducing the average for the lap concerned to a mere 25 mph (40 km/h). Despite this, Pither managed to secure the record with a distance

Above *Stanhope Spencer at Brooklands setting out after records in company with Billy Elce and W.L.T. Rhys. He pushed the '500' one-hour record average up to 65.45 mph.*

Below *Stanhope Spencer and his record-breaking Rudge racer.*

W. Stanhope Spencer who, at the Nottingham MCC's Clipstone Speed Trials on August 19 1911, set a new 500cc solo flying-kilometre record of 72.58 mph.

Frank Pither setting up the first-ever 500cc sidecar record at Brooklands on Tuesday, October 30 1911.

The Rudge racing team at Brooklands: left to right, Stanhope Spencer (No 37), Billy Elce (No 34), Edward Bradford Ware (No 31) and W.L.T. Rhys (No 30).

With the imposition of stricter silencing regulations at Brooklands towards the end of 1911, Stanhope Spencer started experimenting with a longer exhaust pipe terminating in a perforated fishtail silencer.

of 40 miles 1660 yards (65.93 km) in the time.

Not satisfied with his earlier performances, Stanhope Spencer took to the track again on Monday, November 6. The weather was bitterly cold but, despite this and a strong wind blowing throughout the attempt, he managed to take the 500cc solo 150-mile (241.5 km), three-hour and 200-mile (322.1-km) world records at around the 57-mph (92-km/h) mark. Even the wearing of two waistcoats, two sweaters and leathers couldn't keep him warm though, so after three hours he decided to call it a day.

So Rudge-Whitworth's first complete year of motorcycle production ended with a flourish. The firm's products had taken the top sidecar awards in the Leeds-to-London Reliability Trial and the Motor Cycling Club's Reliability Trial, proving them to be not only fast but also absolutely reliable, a great tribute to John Pugh's basic design concept.

As we have said, while all these activities were going on, the new Multi gear (the subject of British Patents 14663/1911 to 22400/1911) was being steadily developed. With this the normal Rudge-Whitworth engine-shaft clutch was retained, but the engine-shaft pulley itself was made in two separate halves. The outer one was attached to the clutch, but the inner one could be moved in or out under the control of a tank-mounted lever which, when operated, engaged with ramps on the outside of the crankcase. Pushing the lever forward moved the pulley up the ramp and pushed it outwards towards the clutch, thereby reducing the space between the two pulley halves. The belt was therefore forced to run at a larger diameter in this space. Moving the lever back to its original position reversed the process.

The control lever also operated a bell-crank lever at the centre of the rear wheel spindle. This device closed up the rear wheel pulley via a short pushrod as the engine-shaft pulley opened, and opened it as the latter closed. The effect of this was not only to keep the driving belt tension constant but, unlike previous designs of variable gear belt drive, to keep its alignment constant as well. It was a really significant development in variable-speed drives.

Relative movement of the outer and inner rear wheel flanges was prevented by interlocking the two components by means of some 76 spokes built into the wheel.

The new 'Multi' gear provided a stepless variation in ratio from a top gear of 3.5:1 down to 7.0:1, using the standard rear wheel, although the top gear could be raised by using a rear wheel with a smaller belt rim. Of course this also affected the lowest gear obtainable as well, since the ratio of top gear to bottom gear, of necessity, remained constant.

The device was tested throughout November and December 1911, without any major development snags emerging, so by the end of the year everything was set for production to commence.

Gearing for success (1912-1913)

Four new 1912 models were exhibited on the Rudge-Whitworth stand at the 1911 Olympia Motorcycle Show in London. One was fitted with the latest version of the Mabon variable gear and had a sidecar attached. The second was a TT model with fixed gear. The remaining two were standard roadsters, but with slight differences in their engine design in that a new type of cylinder barrel was being tried out on one of them. Whereas the standard, already catalogued, 1911 version had a valve chest that was almost circular in plan, the new barrel had deeper finning around the valve chest. The net result was to change the plan view of the cylinder and head from a 'figure 8' to almost an oval. The change had little effect on performance one way or the other, however, and so was later dropped. There were, however, other changes including the increasing of mudguard width from 3½ up to 4 in (89 up to 102 mm) and the addition of a large mud flap to the rear mudguard.

The Rudge-Whitworth catalogue for January 1912 had quite a long introduction to the new Multi gear. The necessary tooling for its production had been completed earlier in the month and the first two production Multi-geared models emerged from the factory on February 3. In the catalogue it was merely mentioned that a Multi-gear version of the standard clutched Rudge could be obtained for an additional £5. The TT model was listed as 'stripped for Brooklands' and ready to race, complete with racing number plates ready to paint, at £47 15s.

In the meantime, unhappy with existing carburettors on the market (particularly for racing purposes), the company had developed a new design based on the scent-spray principle. In this the main jet for the petrol projected into the centre of the induction tract and was fed directly from a float chamber providing a constant head of liquid. The inlet pipe around the jet rotated on a hollow spindle containing the jet, enabling it to act as a rotatable throttle and so cut off the incoming air supply. Prototypes of this so-called 'Senspray' carburettor were first tried on Rudge machines towards the end of 1911 and by the end of January 1912 it was a standard fitting on all Rudge-Whitworth motorcycles. This design was particularly attractive for Brooklands racing since, unlike traditional carburettors of the day, with the throttle lever fully open, the carburettor body rotated to give a virtually unobstructed induction tract.

A more obvious change for 1912 was the decision to put the marque's name on the petrol/oil tank in addition to the three small transfers in use until then, which comprised one stating that Rudge-Whitworth was 'Britain's best bicycle', the Royal Coat of Arms and the legend 'By Appointment to HM the King' — the last deriving from the use of Rudge-Whitworth bicycles by the Royal Family (rather than motorcycles) at Sandringham.

The 1912 Brooklands racing year started off well

The Brooklands Racing Rudge model for 1912, announced in November 1911.

A Rudge rider in trouble on ice at Brooklands Track in January 1912, after the winter repairs had been completed.

D.C. Bolton and his crimson-coloured Rudge, at the Second 1912 BMCRC Brooklands' Race Meeting, after beating the redoubtable George Stanley (Singer) into second place.

for the marque with a win in the 350-500cc solo class of the Hundred-Mile All-Comers' Scratch Race for private owner and future Morgan three-wheeler exponent, Edward Bradford Ware, at the First BMCRC Members' Race Meeting of the year on Wednesday, March 12. He averaged a fine 55.71 mph (89.71 km/h) for the distance. Young 16-year-old John Wallace who made his track début in the same race, had just completed his ninth lap and was approaching the end of the Railway Sraight, when the driving belt of his Rudge broke and jammed between the belt rim and a rear frame member of his machine. His bike went into an uncontrollable skid and flung him off, fortunately on to the soft inside grass verge of the track and without injury. Wallace was to become well-known in the early 1920s as a

A Rudge outfit about to start in the Sidecar Handicap at the First BMCRC 1911 Brooklands' Race Meeting.

rider in sprint events on a Duzmo machine of his own design and eventually became Chief Test Plant Engineer with Napier Aero Engines, a post from which he retired in 1962.

The Brooklands Automobile Racing Club (BARC) staged two motorcycle races at its first race meeting of the year on Easter Monday, April 8 1912. In the first of these, the Short Motorcycle Handicap over 5¾ miles (9.26 km), Rudge rider D.C. Bolton took his crimson-painted machine to a fine win at 62.5 mph (100.6 km/h) after starting 56 seconds before the 'scratch' man. Later the same month, at the second BMCRC race-meeting of the year on Saturday, April 20, he again set up some fine performances. In the Record Time Trials over a flying start on the Railway Straight, he came second to George Stanley (499 Singer), the then reigning Brooklands champion in the 500cc solo class, with speeds of 68.74 and 68.30 mph (110.69 and 109.98 km/h) respectively for the kilometre and mile, while in the three-lap 500cc solo scratch race he surprised everyone by actually beating Stanley. Stanhope Spencer got away first on his Rudge, but by the end of the first lap Stanley had a good lead with Bolton's Rudge coming up fast. By the end of the second lap, they were battling for the lead. At the finish Bolton beat Stanley by barely a length to average an excellent 64.18 mph (103.4 km/h) for the two circuits. Spencer finished third exactly 3 mph (5 km/h) slower.

May 1912 saw the publication of a revised Rudge-Whitworth motorcycle catalogue to take account of all the specification modifications that had taken place during the first few months of the year. The 499cc Rudge-Multi was now more prominent a model in the company's list, with the free-engine and fixed-gear models as optional alternatives by special order. The TT and Brooklands machines were now quite different models and were advertised as such. The Brooklands model had no mudguards and was fitted with a Senspray carburettor and 'straight-through' exhaust pipe. The TT machines on the other hand, included mudguards, a toolbox and various road racing extras — a 'TT Replica' in all but name despite the fact that Rudge-Whitworth machines had as yet gained no Isle of

Top left *N.O. Soresby (Rudge) after winning the 560cc Handicap at the MCC's Brookland's Race Meeting in July 1912.*
Above left *The ever-smiling Billy Elce at Brooklands in 1912.*
Left *Stanhope Spencer starting a successful attack on the 500cc sidecar 2-hour record.*

Man honours; it cost £48 15s. Amongst the extras available for the serious racing man, were a larger combined petrol and oil tank fitted with quick-filler caps, giving a petrol capacity of 2 gallons (9.1 litres) and able to house 4 pints (2.3 litres) of oil, knee grips and a cable-operated magneto control for handlebar mounting. To allow for the operation of the Multi gear, the rear brake had now been changed to a curved shoe which pressed on to the outside of the belt rim when in operation. Also, from June 1912 onwards, beaded-edge tyres were fitted as standard items.

For the past two years, sidecars had become more and more popular with touring motorcyclists and John Pugh realised that there was a lucrative market to be tapped in their sales. Unfortunately the 3½ hp Rudge-Whitworth was at a power disadvantage here compared with 8 hp twins of the day. The standard Multi gear provided the required flexibility of road performance necessary when pulling a sidecar, but it might not respond satisfactorily to the doubling of engine power resulting from the introduction of a new 8 hp twin Rudge engine specifically for sidecar work. John Pugh decided that a compromise solution was required — a larger capacity single-cylinder power unit. Plans were drawn up for a 749cc engine, based on the existing 499cc power unit, with the stroke lengthened from 88 to 132 mm. This involved the fitting of a longer connecting rod and cylinder barrel, and chamfering the mouth of the crankcase to permit adequate clearance between it and the rod during its angular sweep. Heavier flywheels and a longer inlet pushrod were also fitted while the cut-out in the tank had to be increased to accommodate the greater engine height.

The 1912 Isle of Man TT Races were nearly not held at all. After the previous year's event, there had been many complaints about the allegedly 'irresponsible and bad behaviour' of certain riders. There was talk of transferring the races to France or Ireland. Several dealers and manufacturers, including Rudge-Whitworth, signed an agreement not to enter the 1912 event because they felt that the race had departed too far from its original intention of fostering the development of touring motorcycles and because of the flagrant breaking of, for example, silencer regulations the previous year without the culprits being brought to book — no doubt they had in mind the somewhat primitive exhausts used on the racing Indian machines. The new ACU secretary, Tom Loughborough, had as a result a 'mess of porridge' to clear up. In the end, a TT was held in the Isle of Man but without an official Rudge-Whitworth team entering. So there were no successes for Rudge machines in 'The Island', in 1912.

It was around this time that a new craze had appeared on the scene, the cyclecar. This was the lightweight three- or four-wheeled car powered by a motorcycle-type engine. A number of firms were busy developing these inexpensive vehicles, while Harry Morgan had been producing a three-wheeled runabout of the type, powered by a transversely-mounted V-twin JAP engine, for the past two years or so. John Vernon Pugh, ever on the look-out for new manufacturing possibilities, persuaded his Board of Directors that this was a good area to investigate. The firm was already involved in wheel, carburettor, bicycle, motorcycle and sidecar production, and was keen to get into new ventures involving similar types of product so, since there was no TT involvement that year, the Summer of 1912 saw John Pugh and Frank Pountney developing a new prototype four-wheeled cyclecar, using the new 749cc single-cylinder engine as its power unit. This was mounted between the front wheels, which were widely splayed and made necessary the knocking down of part of a wall to widen a factory doorway before the car could be wheeled out for its first trial run in the early hours of one cold autumn day in October 1912.

The Standard Rudge-Multi introduced in 1912.

The 500cc version of the Rudge cyclecar made in 1913.

The new two-seater vehicle had standard Rudge-Whitworth motorcycle front wheels. With a standard Rudge clutch fitted to the engine shaft, the power was transmitted to the two Rudge-Multi rear wheels at the back of the vehicle via an enclosed belt to a transversely mounted countershaft under the driving seat. Fixed belt pulleys at the ends of this shaft each drove a rear wheel via external belts. The pulleys on the rear wheels opened and closed in the style of the normal Multi gear, but the fixed pulleys on the countershaft, being eccentrically mounted, could be moved in a small arc to take up belt slack. The system resulted in an overall drive ratio ranging from a top gear of 3.875:1 down to 9.0:1. This with a vehicle dry weight of 504 lb (229 kg).

Engine starting was by handle on the offside of the bonnet, the top of which was formed by a motorcycle-type combined petrol and oil tank mounted above the power unit.

The vehicle's wide track of 4 ft (1.2 m) was said to give it exceptional stability, even when travelling flat out at its maximum speed in top gear. This stability was further aided by the adoption of rack-and-pinion steering, with the car underslung on semi-elliptic leaf springs at the front and quarter elliptic ones at the rear. A very complimentary report on the Rudge cyclecar's performance appeared in a road test of the vehicle published in *Motor Cycling* magazine in October 1912. The first prototype was somewhat ugly though a much neater 500cc version was introduced early in 1913. Neither, so far as I know, though, actually went into production.

Another development for 1913 was the introduction of an improved 'Colonial' model able to cope with the rougher and usually unmade roads of British territories overseas. This was fitted with an especially robust frame since a number of frame breakages had occurred, particularly with machines exported to South Africa, using the standard Rudge frame. This was introduced in September 1912 for the coming year. At this time almost half the factory's annual motorcycle production was being exported, a large proportion being sent to the Rudge-Whitworth depot in Johannesburg.

The firm, as mentioned, was also taking an active interest in sidecars at this time and, on November 28 1912, an experimental sidecar outfit was taken down to Brooklands for extensive testing. It had an underslung chassis with its principal members 4 in (101 mm) below the level of the sidecar wheel spindle. This wheel was interchangeable with the two motorcycle wheels and was of the detachable, car type then being considered for the experimental cyclecar project. Suspension was by means of two coil springs at the front and two at the rear using swinging shackles. With successful testing out of the way, the first production model came out of the factory on December 7 1912, complete with a body which, unlike the then current practice, was not made of wickerwork but was constructed from aluminium panels mounted on an ash frame. Of streamlined shape, it had the overall appearance of a fairground swing. It was fitted with a small access door and had a small fibre foot mat for the passenger. Finished in the same colour, green with a black lining, as the Rudge-Whitworth motorcycle tank, it was priced at £16 complete with chassis and for a further £5 could be obtained fitted with a bracket for a sidecar lamp as an optional extra. By February 1913, it was in full-scale production at the rate of 30 a week and was soon being seriously considered by Brooklands' racing men for use in the new sidecar races at the track.

A team of four riders, comprising Cyril Pullin, Frank Bateman, H. Hill and S. Heales, was entered for the first Brooklands race-meeting of the year held on Easter Monday, March 24. With F. Clark, the Works' Foreman Tester, in charge and John Pugh himself in attendance, they were obviously under pressure to do well. Things started promisingly in the first motorcycle event of this BARC meeting, the Short Motorcycle Handicap run over two laps, with Frank Bateman coming home second after getting away 46 seconds ahead of the scratchman. But things really went the Rudge way in the 8½-mile (13.7 km) three-lap Long Motorcycle Handicap, Cyril Pullin leading all the way from the end of the first lap to the finish. He averaged a good 60.3 mph (97.1 km/h) and was followed home by Bateman and Hill who filled second and third places. A fine one-two-three finish.

At the postponed First 1913 BMCRC Brooklands Race Meeting on Wednesday, April 2, Rudge riders really shone. All the races were run over three laps. George Stanley dominated the 500cc solo Scratch Race, winning it at 67.94 mph (109.4 km/h), but Rudges in the hands of Frank Bateman and S. Heales were a good second and third. The new 499cc Rudge outfit, driven by L. Hill, gained a useful third place in the 1000cc Sidecar Scratch event. In the 500cc solo Handicap, Stanley was not so lucky, Vernon Taylor coming home first on his single-geared Rudge from a 54 seconds' start over scratch, at a fine average speed of 64.46 mph (103.8 km/h). The new Brooklands' Rudges were clearly proving extremely competitive.

In the Sidecar Handicap, scratchman Freddie Barnes on his big 986cc Zenith outfit had all his work cut out trying to overhaul L. Hill's 499cc Rudge combination, which proved faster even than A.J. McDonagh's big 749cc single-cylinder Rudge outfit in this event. Barnes only just succeeded in overtaking Hill before the finish.

In the 1000cc Solo Handicap, P.F. Glover took his Rudge-Multi home to the first-ever recorded victory for a multi-geared Rudge at Brooklands. Up to that time the Multi gear had not been much used at the track, regular Rudge-riding Brooklands men preferring single-geared machines for track work, since it largely involved full-throttle operation. Glover averaged 56.29 mph (90.64 km/h) for the three laps of the race.

Instead of returning to the Rudge-Whitworth works in Coventry after this meeting, L. Hill and Frank Bateman stayed on at Brooklands and the following Thursday, set about attacking 500cc sidecar world records. With Hill driving and Bateman in the 'chair', the duo set up new speeds of 53.77 and 54.22 mph (86.59 and 87.31 km/h) for the flying-start kilometre and mile respectively.

At the Second BMCRC Brooklands Race Meeting on April 26 1913 were run the Junior and Senior Brooklands TT Races, to enable potential entrants in the forthcoming Isle of Man races to test their machines in a tough long-distance event. In the 70-mile 1161-yard (113.78-km) Senior race there were 16 starters including L. Hill, S. Heales and Frank Bateman on their Rudges. Hill led at the outset, followed by Heales and George Stanley (499 Singer), the eventual winner, who was followed in turn by Bateman in fourth place. Hill lost the lead to Stanley on the second circuit, but regained it on the fifth, followed then by Bateman in second spot. Lap 14 saw Hill drop out of the running altogether, as did Bateman after being overtaken for the lead by Stanley who went on to win at 57.82 mph (93.11 km/h). The only other Rudge rider to finish, S. Heales, came home third behind Stanley's team mate Victor Horsman (Singer) at 55.55 mph (89.45 km/h).

This sidecar was standardised for 1913.

P.F. Glover at the BMCRC's April Brookland's Race Meeting in 1913 after becoming the first rider to win at the track on a Rudge-Multi.

At the BARC Brooklands Race Meeting on the Whit Monday, A.J. McDonagh (499 single-geared Rudge) easily won the two-lap (9.2-km) Short Handicap with only 34 seconds start from scratch, at the very fine average speed of 65.8 mph (109.96 km/h). Later the same day, in the three-lap (13.7 km) Long Handicap, Frank Bateman, using the Multi gear of his Rudge to great advantage, scored a win at the even higher speed of 67.5 mph (108.70 mph).

The following Saturday at Brooklands, the Rudge team was once again out in force, this time on the occasion of the Third 1913 BMCRC Members' Race Meeting. Proceedings commenced with the holding of the Record Time Trials in which Frank Bateman, riding the Rudge-Multi on which he was entered in the forthcoming Isle of Man Senior TT Race, recorded speeds of 74.57 mph (120.08 km/h) over the flying kilometre and 73.47 mph (118.31 km/h) over the flying mile. It could have been the attainment of these high speeds that contributed to his downfall in the Senior Brooklands TT Race that followed and in which he retired with tyre trouble. In fact both A.J. McDonagh and L. Hill had trouble in this race and were forced to retire. Only S. Heales amongst the original Rudge entries completed the event and he finished in second spot at an average of 60.77 mph (97.86 km/h) some 2.6 mph (4.2 km/h) slower than the winner of the race George Stanley (Singer).

The 749cc big-single Rudge sidecar outfit made a re-appearance at the track on Saturday, May 24, at the Essex Motor Club's Race Meeting. In the hands of G.T. Gray it scored its first race victory by winning the two-lap Sidecar Handicap over 5¾ miles (9.2 km) at 47.5 mph (76.5 KM/h) average.

While all this track racing had been going on, a team of British Rudge riders under the control of F.A. ('Rowley') Rowlandson, the Rudge concern's Competition Manager, had been gaining some true road-racing experience on the Continent prior to the Isle of Man races. Thus, on April 26 and 27 1913, an official Rudge works team comprising Rowlandson, Scott and Newman, had taken part in the gruelling 683-mile (1100-km) International Road Race in Italy. Run over a contorted route consisting principally of dusty unmade tracks, the course passed through Turin, Brescia and Bologna. At the end of the first day of the event, from Milan to Turin and Bologna, Rowlandson led with the Italian Rudge rider Vailati in third place. Only 20 out of the original 60 starters survived to make the return journey, which commenced at 4 am. At Levo, five hours later, Rowlandson crashed, but managed to get going again to finish second, despite his wheels being found to be out of line by some 8 inches (20 cm)! Before his crash he had also taken part in and had won the Aprica Hill Climb in which he was 1 min 30 sec faster than any other rider thanks to the deft use of his Multi gear. In the main race, only 14 of the original competitors crossed the finishing line, Vailati being the overall winner at an average speed of 30 mph (48 km/h) an hour ahead of Rowlandson. All in all an excellent debut for

A.J. McDonagh, after winning the Short Motor Cycle Handicap at the BARC's Whit-Monday Brookland's Race Meeting in 1913.

Rudge-Whitworth in Continental road racing.

Rowley stayed on in Italy after the Milan-Bologna event to take part in a race on the Cremonese Circuit on May 18. A three-lap 65-mile event, this race ran through the village of St Giovanni and Piadenna, and gave him a fine win in the 500cc class at an average speed of 55 mph (88.6 km/h). The crude organisation and conditions of such events in those days is reflected by Rowley's subsequent memory of that race as 'involving the wholesale slaughter of chickens, which strayed on to the circuit to be caught unawares and to be cut off in their prime by impact with motorcycles travelling at over 50 mph (80 km/h)'. The Rudge-Whitworth team, comprising Rowlandson, Vailati and Fossati, also took the Manufacturers' Team Prize in this event.

Whilst all this Continental activity had been in progress, back in England over better roads, ten Rudge riders had contrived to get Gold Medals in the 800-mile (1288-km) long London-to-Edinburgh Trial organised by the Motor Cycling Club. But now the attention of the British motorcycling fraternity was being concentrated upon the forthcoming TT Races in the Isle of Man. The Rudge-Whitworth concern had entered a strong team in the Senior event and had high hopes of victory in view of its recent widespread racing successes. Certainly a TT win would work wonders for sales.

The 1913 Isle of Man Tourist Trophy Races were unique in that they were run in two parts on different days. The plan was to have two separate races, the Junior for the '350s' and the Senior for the '500s', run one after the other over three laps on Wednesday, June 4, and on the Friday to send the survivors off in one final four-lap race to decide the winners. This method was adopted to cope with the amazingly large entry of 147 machines and their riders — 103 '500s', including 12 Rudges, and 44 '350s'. To distinguish between them, the Senior riders wore red coloured waistcoats, while the Junior men had waistcoats coloured blue. A special Manufacturers' Team Award was also instituted, which cleverly thwarted any question of a trade boycott of the races, such as had occurred the previous year.

Hugh Mason (347 NUT), despite coming straight from hospital after an injury sustained in practice, rode successfully to win the Wednesday morning Junior qualifying race. In the afternoon, 97 Senior machines faced the starter. At the end of the first lap, Rudges held the first four race positions and Frank Bateman led on the road. Indian machines were fifth and sixth, followed by a Matchless, a Scott and the rest of the field. The much vaunted Scotts that had done so well the previous year, were well and truly out of the running at this stage in the proceedings. But H.O. 'Tim' Wood (Scott) put on a spurt and by the end of the second circuit had forced himself into second place. The end of the third and final circuit saw him come home first ahead of Frank Bateman who finished second on his Rudge-Multi.

Following a day's rest, the surviving 47 Senior

G.T. Gray, who rode this 750 Rudge outfit to Victory at the Essex Motor Club's Brooklands' Race Meet in May 1913.

riders and 28 Junior men lined up for the four-lap composite final on Friday, June 6. They were sent off alternately at 30-second intervals, first a 500, then a 350 and so on. Now in those days, the Isle of Man circuit more closely resembled a dust-laden cart track in parts than a racing circuit. There were many loose stones about and one of these, thrown up by another machine's rear tyre, pierced the hose of the radiator on Tim Wood's water-cooled Scott on the very first lap. He was forced to stop and make hasty repairs. This put Frank Bateman in the lead, followed by Alfie Alexander on the Indian and Ray Abbott on another Rudge. Then, on the second lap, Alexander took the lead, which was regained by

Frank Bateman at Brooklands with his Rudge-Multi, a month before the 1913 TT Races in which he was killed.

P.H.A. Matthews, who won the 500cc Class of the 1913 Brooklands' Race Meeting of the Public Schools MCC.

Bateman on the climb up the Snaefell mountain road. Coming down the other side of Snaefell, disaster struck. In the lead, Frank Bateman's machine got caught in a rut at the side of the road, while rounding Keppel Gate corner close to the 34th milestone, which pulled the beaded-edge front tyre right off its rim. Crashing into the bank at the side of the road at something approaching 70 mph (113 km/h) he sustained injuries from which he subsequently died in hospital.

This left Alexander in the lead, but not for long since carburettor trouble was slowing him and by the end of Lap 2 Ray Abbott became the new race leader. The last lap saw a tense drama develop between the three leading contenders in the race. Coming over the Mountain for the last time Abbott held a lead of about 4 min which was diminishing rapidly in favour of Tim Wood who had overtaken Alexander (Indian) and was riding magnificently. Then Abbott on the Rudge-Multi threw it all away by overshooting the corner at St Ninian's near Bray Hill and having to take the slip-road as a result. He stalled his engine in the process and had to restart again, which wasted further valuable time.

Tim Wood, who had also run out of road at one stage, had managed to keep *his* engine running. It was clearly going to be a close run thing! Ray Abbott finished the race still not knowing whether his time was fast enough to give him victory or not, since he had started some 1½ min ahead of Wood and the Scott rider had not yet finished. The result would now depend upon whether or not Tim made any last minute mistakes. It must have been an agonising wait for him too! But then came the two big disappointments — he had lost, but only by 5 seconds! If only he had not stalled his engine, Rudge-Whitworth might have gained its first Senior TT victory.

In later years, when asked how he came to make such a mistake and on his last lap too, Abbott explained that he had already been in the saddle some 2½ hours on his rigid-frame machine, riding at speeds up to 70 mph (113 km/h) over very rough badly rutted cart tracks for most of the way, so that by the time he had reached St Ninian's corner on his last lap he was close to both physical and mental exhaustion. He was not a local man and did not know the course intimately; so he made the first of his two crucial errors. The second, which also resulted from his exhausted state, was that when he braked he forgot to move the control lever of his Multi gear from its 'high' to its 'low' position. When he restarted, therefore, he was attempting to get underway in top gear and promptly stalled his motor. The time taken to restart and get back into the race, of course, was considerably longer than the 5 seconds by which he lost it. But he should have had a greater safety margin on time than this anyway, and it was considered that the need to take the hand off the handlebar every 5 miles or so to work the tank-mounted manually operated oil pump was a contributory factor to his losing the

B. Pattison (left) and C. Meredith (right) on their Rudges at the RAC's 1913 Brooklands Gala Race Meeting.

race. The need to operate this pump, as well as the Multi gear, and steer the machine safely over the inadequate road surfaces of those days placed just too much strain on the rider in a long race like the Isle of Man Senior TT. Clearly something would have to be done to remedy the situation before the following year's races came round.

It was John Pugh, with his usual inventiveness, who came up with the answer. He realised that the right foot was relatively unoccupied throughout the race, so he designed a foot-operated oil pump, a foot-pedal-actuated, tube-and-plunger type of pump, housed inside a separate oil tank located beneath the saddle, the main tank now being confined solely to petrol. The plunger of this pump was kept at the bottom of the tube by a strong spring which, together with a ball valve, prevented the oil running down into the crankcase uncontrolled. With the pedal depressed, the plunger forced a charge of oil into the rear of the crankcase. After a period of testing in competitions during the remainder of 1913, the system became part of the standard production machine's specification for the coming season.

In order to recoup some of the prestige lost by Rudge-Whitworth over the Senior TT result, Team Manager Rowlandson took a works' team over to France to take part in the Motor Cycle Grand Prix to be run on the Circuit de Picardie on July 13. Irishman Tommy Greene, who had had trouble in the Isle of Man race, took the lead in this event at its half-way stage and held it to the end. The other Rudge teamsters finished seventh and eighth respectively in the hands of Abbott and Rowlandson. Well known for his Rudge successes in Ireland, this was Tommy's first international race win.

Despite the firm's inexorably bad luck in the TT, throughout the rest of 1913 the number of sporting successes of its products mushroomed and by November that year more than 300 first places had accrued to the marque. There were wins in such major events as the Grand Prix of Denmark, the Championship of Russia, the South African Continental Cup and the Open Championship of Belgium. In the Second International Motorcycle Trial between England and Holland, Ray Abbott was a member of the British Team, while two of the Dutch team members, Citroen and Neudenberg, also rode Rudge-Whitworth machines.

At Brooklands, it was Cyril Pullin who kept the flag flying for the Rudge marque during the long summer of 1913. His first success came exactly a fortnight after the TT Races in the Isle of Man, at the BARC's Mid-Summer Race Meeting held on Saturday, June 21. This was in the two-lap Fifteenth Short Motorcycle Handicap in which he rode in company with Rudge team-mate Vernon March. Moving up fast he had taken the lead by the start of the finishing straight on the last lap, to win at **66.5 mph (107.8 km/h)**. Unfortunately, March blew his engine up badly and came into the pits with his cylinder and parts of his crankcase missing.

In the 8½-mile (13.7-km) Benzole Handicap, sponsored by the *Daily Express* newspaper, in which all competing machines had to run on Benzole as the fuel, George Stanley proved just too

Cyril Pullin (single-geared Rudge), after winning the two-lap motorcycle handicap at the BARC's October Race Meeting of 1913.

fast on his 499cc Singer and Cyril had to be content with second place at the finish. His next outing at the track was at the annual MCC Brooklands Race Meeting. Other Rudge riders present that day included N. O. Soresby and Rowley Rowlandson. Unfortunately there were no wins for any of them on that occasion.

Excitement and anticipation had been building up for weeks in anticipation of the marathon Six-Hour Multi-Class Scratch Race which the BMCRC intended to run at its Fifth 1913 Race Meeting at Brooklands, to be held on Wednesday, July 16. The race had been organised to give riders of machines of all capacities the chance of breaking records in their classes. It was split into eight separate races to be run simultaneously for 350, 500 750 and 1000cc solos, 500, 750 and 1000cc sidecar outfits or three-wheeled cyclecars, and 750 and 1000cc four-wheeled cyclecars.

Cyril Pullin was entered in the 500cc solo class on his 499cc Rudge Multi. He was the first to complete 50 laps and at that stage led his class. After three hours' running, however, the earlier leader, Jack Haswell (Triumph), had again nudged ahead of Cyril, both having completed 68 laps. The former maintained his lead to win at 58.62 mph (94.40 km/h) with Cyril Pullin second at 55.82 mph (89.89 km/h).

L. Hill and A. J. McDonagh were the principal Rudge-riding contenders at Brooklands during the BMCRC's Sixth 1913 track meeting on Saturday, August 9. The races were all over 10 miles, starting on the Railway Straight and finishing at the Fork, and in the 500cc event, Hill scraped home second behind the redoubtable George Stanley (499cc Singer), at an average speed of 63.62 mph (102.44 km/h) just 0.6 seconds ahead of Victor Horsman (down in the programme as 'E. H. Victor'), who was also on a Singer. The weather throughout the meeting was dull and cool, most unlike the rest of that summer.

Fortunately the weather was fine and sunny for the BMCRC's seventh monthly meeting the following month and, despite a strong north-westerly wind, speeds were high. But luck seemed to be out for L. Hill and the other Rudge team members, who suffered persistent driving-belt trouble. Fortunately such troubles did not afflict Cyril Pullin's Rudge-Multi at the BARC Brooklands Race Meeting on Saturday, October 4 1913 and, in the first of the two motorcycle races on the programme, the Seventeenth Short Handicap, a two-lap affair over 5¾ miles (9.26 km), he went like the wind. He passed Cyril Williams on his fleet 350 AJS at the start of the finishing straight near the end of the last lap and sprinted home to a well judged win at 68.8 mph (110.79 km/h), the highest race speed yet for a Rudge-Whitworth at the track. The heavens then opened and the rest of the meeting had to be abandoned due to the rain.

As if to make up for the appallingly bad weather of the previous two weeks, that on Saturday, October 18, the occasion of the BMCRC's Championships at Brooklands, was almost spring-like. Cyril Pullin had carried out a number of modifications to his machine for the meeting and his Rudge now sported a large rear-mounted oil tank encircling the rear down tube of the frame in the manner of later machines. He had entered his machine, with sidecar attached, in the Multi-Class Hour Race for Sidecars and Cyclecars. Teddy Ware (748cc Zenith-Gradua-JAP), Charlie Collier (998cc Matchless-Gradua-JAP) and P. H. T. Hoare (986cc C & H-JAP) held the first three places in the race until Lap 7. Pullin seized third spot on Lap 9 and this order was maintained until the end. He won his class (500cc sidecars) to set a new world record for the hour of 52 miles 764 yards (84.43 km).

Quickly unbolting the sidecar, he got his machine ready for the next event in which he was entered, the Senior One-Hour ACU Championship for *The Motor Cycle* Challenge Cup, which was for solos up to 500cc. He, together with L. Hill and Tommy Greene, made up a strong Rudge contingent in this event. George Stanley (Singer) did not run, but his stable-mate Jimmy Cocker did and led at the end of the first lap, with Hill second and Pullin third. Hill displaced Cocker on Lap 2, then on Lap 3 Pullin shot into the lead and stayed there for the next five laps. Dan O'Donovan's Norton posed a threat for a while until its engine seized on Lap 10 when in second spot. Meanwhile Pullin stayed in front and very little distance separated the next two men, Cocker and Hill.

Above *One of the faster Rudge riders at Brooklands in 1913 — Vernon March.*

Below *Cyril Pullin (right) and Tommy Greene (left), have just finished first and third respectively in the ACU's Senior One-Hour 1913 Brooklands' Championship Race for* The Motor Cycle *Cup.*

By the end of 15 laps the third man, Hill, had made two stops and Irishman Tommy Greene was breathing down his neck. Two laps later Pullin stopped for oil and Cocker took the lead, holding it for four laps. Pullin got away well after his stop and was soon catching up his rival. Then, on Lap 22, with everyone straining to see who would be leading as the hour approached, Pullin gradually closed the gap and passed Cocker. He won on time as well as on the track covering 64 miles 289 yards (103.32 km) in the hour.

As was usual, just before the annual Motorcycle Show, riders were out on Brooklands Track after records. The Rudge team was determined not to miss out on any potential record spree and on Thursday, November 20, Tommy Greene and Cyril Pullin, riding alternately, made a successful onslaught on Class C (500cc solo), Class D (750cc solo) and Class E (1000cc solo) world records from five to seven hours and from 250 to 400 miles, at average speeds of just under 60 mph (97 km/h).

Cyril Pullin, with friends, after a successful bout of record-breaking at Brooklands in November 1913.

4

Experiments, prototypes and winners (1914)

Just before World War 1, several motorcycle designers had realised that there was a market for an economical lightweight motorcycle of under 300cc capacity. A number of firms, including Ariel and Triumph, were developing prototypes of such models during 1913 and 1914, as well as Rudge-Whitworth.

It was in April 1914 that the first prototype of the new 275cc Rudge appeared. It was powered by an overhead-valve (ohv) single-cylinder four-stroke engine fitted with an outside flywheel and employing a most unusual form of valve gear. This involved vertical parallel valves with a beam mounted over them and pivoted from the centre of the cylinder, which was actuated by pushrod. This pushrod was operated by a cam follower shaped like a large letter 'C', which was positioned around a single-lobed camshaft. With the pushrod raised, the beam rocked and opened the inlet valve. The exhaust valve was opened by the return of the pushrod when the cam lobe was at its lowest position.

To allow for thermal expansion, the pushrod was fitted with a conventional tappet. To enable the exhaust valve to be opened, a spring was fitted on to the pushrod to force it down. Thus, on the induction stroke of the engine, the cam had to act against two springs, causing excessive cam wear. Later in April, the engine was redesigned on more conventional Rudge-Whitworth motorcycle lines, with an overhead inlet valve and side exhaust valve. Unfortunately, even in this form it did not go into production due to the forthcoming war.

For 1914, the ACU reverted to its usual one-day/one-race system for the Isle of Man TT Races. The two-day races of 1913 had been introduced largely to test restarting from cold after the first day's laps. Very few riders had experienced any trouble though, on that score, and it was decided that the extra complication and expense of the two-day system just wasn't worth the trouble it entailed. The previous year (1913), however, the majority of machines had shown that they were reliable enough for the distance, so for 1914 the ACU split the difference between the four and five laps for the Junior and Senior Races of 1912 and the six and seven laps (*in toto*) of the 1913 events, with a Junior Race of five laps and a six-lap Senior event. Crash helmets (now called 'safety helmets') of ACU-approved type were made compulsory, as were also security bolts for beaded-edge tyres, in view of Frank Bateman's fatality of the previous year. Also, riders had to complete at least six qualifying laps during the practising periods, to ensure familiarity with the course.

Six Rudges were entered in the Senior Race, scheduled for May 21, ridden by J.W. Adamson,

The Rudge team hard at work 'tuning' prior to the 1914 Isle of Man Senior TT Race.

Cyril Pullin (right) and other Rudge teamsters posing for a Rudge-Whitworth publicity photograph during the TT Race period.

J.W. Adamson (Rudge) who, but for tyre trouble, might well have provided Cyril Pullin with serious competition for 1914 Senior TT Race honours.

Rudge rider J. McMeekin (No 123) at the start of the 1914 Senior Isle of Man TT Race.

Irish rider Tommy Greene (No 81) about to push his 'works' Rudge over the starting grid as the flag drops at the beginning of the 1914 Senior TT Race in the Isle of Man.

Tommy Greene, P.H.A. Matthews, J. McMeekin, Cyril Pullin and Tom Sheard.

It was at the end of 1913 at Brooklands that 'Rowley' Rowlandson, the Rudge Competition Manager, suggested to Cyril Pullin that he should ride for the firm in the following year's Island event. Cyril was not at all enthusiastic at first. He was primarily a track racer and had little first-hand knowledge of the handling of the Rudge-Multi under road-racing conditions. At Brooklands Track he had mainly used single-geared solo Rudges since there was little need for gear changing once the machine was flat out on the Outer Circuit. He had, however, ridden in the 1913 Junior TT on a Veloce, the forerunner of the Velocette, and had formed the conclusion that the TT Rudge-Multis of that year were certainly fast, but left a lot to be desired on the handling front. He finally agreed to ride providing a machine was built to his own pattern. Rowlandson and John Pugh agreed to this proposal, so matters proceeded from there.

Cyril Pullin produced a mock-up frame at his workshop in Vauxhall Bridge Road, London, and soon was able to draw up a specification based upon which the Rudge-Whitworth works quickly built some machines incorporating the Multi gear. Extensive tests were then carried out on them, both at Brooklands and on the road, to assess their suitability for the Isle of Man Races. The 1914 TT Rudge Multi turned out to be, in Cyril's own words: '. . . the best steering motorcycle I have ever owned or ridden'.

In April 1914, a month before the Island event, Cyril, 'Rowley' Rowlandson and Tommy Greene took themselves off to Italy to test the new TT machines in the *Giro d'Italia*, a road race around Italy. In those days such continental road races were run over rough tracks, made-up roads being a rarity, and so provided a really tough test. Then it was back to England and the Isle of Man for the start of TT practising.

The start, grandstand and paddock, had been moved to the top of Bray Hill on ground now occupied by Douglas High School. The 'Four-Inch' course was still being used, so the grandstand spectators saw little of the racing since the riders promptly disappeared down Bray Hill after cautiously rounding the corner at Parkfield crossroads. However, the refuelling and repair depots were now located, as they are today, right in front of the stands opposite the race scoreboard. They were formed from one long trench, separated into cubicles which came to be known as 'pits'.

In the Senior Race the traditional struggle between Indian, Matchless, Rudge and Scott teams developed. Tim Wood, on his Scott, led off with a terrific lap in 42 min 16 sec at a speed of 53.50 mph (86.15 km/h) followed by Adamson (Rudge), 2 min behind. Tim increased his lead to nearly 3 min by the end of the second lap, Adamson having dropped back with a piece of broken horeshoe lodged in his tyre. This let Harry Collier (Matchless) into second spot.

On Lap 3 Tim's Scott started to develop magneto trouble, letting Collier into the lead, with Cyril Pullin running third. The next circuit saw Tim in the lead once more only to drop out with a burnt-out magneto. Harry Collier then crashed at over 70 mph (112 km/h), splintering his kneecap and letting Cyril into the lead. The other two Matchless men in the race experienced gear trouble. Charlie Collier stripped the pinions in his hub gear, while Bert Colver had the same malaise while attempting to catch the flying Rudge.

The result was a win for Cyril Pullin on his Rudge-Multi at an average speed of 49.49 mph (79.69 km/h), a mile an hour faster than the previous year's winning average despite having his goggles smashed by a flying stone halfway through the race and having to have the ACU doctor remove some 36 pieces of Manx road metal fragments from his eyes after the race as a result.

Pullin's TT-winning machine of 1914 was able to lap Brooklands Track at a regular 75 mph (120 km/h) and was timed unofficially at 83 mph (133.7 km/h) over the flying-start half-mile at the end of the Railway Straight. The short exhaust-pipe used discharged into a cylindrical silencer box having a series of tangential discharge pipes.

Interesting details of the engine of Cyril Pullin's TT-winning Rudge are revealed in a workshop notebook he maintained when preparing the machine for the race. It differed from standard in several important respects. Firstly, it had a higher compression ratio (4:1) and five per cent nickel steel inlet valve with a shrouded neck. The cams were ground to give an inlet opening at 5° BTDC and closing at 37° ABDC. The exhaust opened at 47° BBDC and closed at 20° ATDC, giving some 25° of overlap. A lightened connecting rod carried needle roller big end bearings, but a plain small end bush. The cast-iron piston had also been carefully lightened and two ⅛-in (3.2-mm) deep rings fitted. Piston clearances were almost twice those recommended by other tuners of the day. The flywheels use were polished and loaded at the rim to balance 45 per cent of the reciprocating mass.

Careful attention had been paid to the adjustment of the Multi gear on this machine to obtain the best possible results. The John Bull driving belt used had been pre-stretched during the practice periods and prior to the race was left to hang up with a 56-lb (25.5-kg) weight attached. As with TT machines in previous years, ultra-large oil and petrol filler caps were employed, but for 1914 the oil tank was a separate affair as mentioned earlier, with a pump operated by the rider's heel.

For the race, Cyril Pullin wore a specially made black-leather riding suit based on the design of a conventional boiler suit, but tightly fitting to cut down air resistance.

An ignition-control lever was mounted on the top of the petrol tank to help negotiation of the slower parts of the course, such as the Ramsey hairpin and the Gooseneck, without stalling the engine, in view of the 7:1 bottom gear employed. The petrol tanks on the other Rudge team members' machines were fitted with knee grips, but Cyril dispensed with these in favour of pads fitted inside the legs of his racing leathers.

Anyway, his overall thinking proved highly successful and he certainly rode the fastest Rudge machine in the race, having tuned the engine himself using knowledge gained at Brooklands. In fact, after the official ACU machine examiners had measured the capacity of the stripped engine of his machine at the end of the race, 'Rowley' Rowlandson wanted to whip his engine back to Coventry as quickly as possible, but Cyril Pullin stopped that ploy by taking his machine straight back to Brooklands. He was not going to reveal his hard-won tuning secrets gratis to anyone.

The other Rudge machines were some way behind Pullin at the finish. J.W. Anderson, who might have provided Cyril Pullin with some opposition dropped back after problems and finished 41st. P.H.A. Matthews came home two places behind Anderson, while H. McMeekin finished in 38th position. It was the local rider, Tom Sheard, who rode the second Rudge home, in 30th spot. The only Rudge rider to retire, though, was Tommy Greene, who had the misfortune to have his petrol tank split open and his machine catch fire as a result at Quarter Bridge.

Some of the track testing of the TT Rudges before the Island event was placed in the hands of L. Hill. He rode one of these, before the TT Races, at the BMCRC Members' Brooklands Race Meeting on Saturday, May 2 1914. In the Record Time Trials that day with sidecar attached, he recorded speeds of 56.49 and 54.05 mph (90.97 and 87.04 km/h) over the flying-start kilometre and mile respectively. Later he was about to move off to the starting line to take part in the 500cc Sidecar Sprint, when he discovered that his oil tank was empty. Oil was soon brought and he made a start, only to break a valve rocker out on the track. However, one of his team mates, G.H. Fry, managed to win the event at 47.23 mph (76.05 km/h).

The works' Rudges returned to Brooklands at the first major post-TT race meeting to be staged by the BARC, on Whit Monday, June 1 1914. Cyril Pullin and his brother 'A.L.' were entered, together with L. Hill, but it was only the latter who gained any race honours that day, finishing third in the 19th Short Motorcycle Handicap over two laps.

Exactly five weeks after the Isle of Man Senior TT Race the spark that was to set alight World War

Cyril Pullin after winning the 1914 Senior TT Race.

I took place, with the assassination on Sunday, June 28, of Archduke Franz Ferdinand and his wife, at Sarajevo in what is now Jugoslavia. It took some time for the importance of this event to register with the British public and, meanwhile, things went on very much as before. When war was eventually declared, on August 4, the Rudge-Whitworth Board of Directors must have wondered whether it would be able to capitalise on its overdue TT success, which appeared to have come too late. Quite apart from that, the new 1915 Rudges had been introduced to the public only a month before — clearly a sales disaster was in the offing.

Nevertheless, one hoped-for sales success of the year had been the introduction of 499 and 749cc models fitted with Sturmey-Archer three-speed gears. These had been introduced back in March and some 100 were produced during the course of that year (1914). Unfortunately, they never proved excessively popular with the buying public since the gears proved a constant source of trouble to riders.

The increasing interest of the motorcycling public in sidecar combinations caused John Pugh to consider production of a big-twin machine. His investigations of their market potential started towards the end of 1913 and he decided to adopt the simplest and most cost-effective approach to exploring the technical possibilities. He took two standard production 499cc Rudge-Whitworth power units and, with their crankshafts joined by a splined sleeve, bolted their crankcases together to produce an experimental 998cc vertical twin. A standard production 499cc Rudge-Multi with sidecar attached, had in effect a second engine grafted on to its timing side, supported by means of additional frame tubes attached to the steering head and saddle lug. An ugly but extremely effective way for him to check his theories!

The use of a splined coupling enabled the two crankshafts to have any desired relative phasing. With the pedals removed, push starting had to be used to get the outfit moving, but with the use of exhaust-valve lifters this did not present insuperable problems and anyway it was never pretended that the machine was anything other than experimental. Each engine had its own magneto and carburettor. A single carburettor was tried, but performance suffered accordingly.

A further experimental vertical twin was produced using two 749cc single-cylinder engines, which had a very healthy performance according to the factory tester. Unfortunately the width of both power units and the addition of more engine weight to the offside of the machine greatly impaired the roadholding of the machine with the narrow-section tyres then fitted as standard on motorcycles.

However, based on this research, a prototype 998cc 60° V-twin Rudge-Multi was produced, which first ran on September 2 1914. It had a pair of cylinders with the customary Rudge inlet-over-

Cyril Pullin (left) flat out on his Rudge at Brooklands Track in 1914, demonstrating his mastery of the art of 'getting down to it'.

Winner of the 500cc one-lap sidecar, G.H. Fry and his Rudge outfit, at the BMCRC's May 1914 Brooklands' Race Meeting.

exhaust valve arrangement, but with the inlet ports pointing towards the inside of the 'V' of the cylinders and fed by a single Senspray carburettor using a T-junction manifold. A second and third prototype appeared on November 3 and 21 respectively as experimentation continued into the winter of 1914-15.

While all this development work was going on, motorcycle production had continued fairly steadily throughout the year at the rate of around 600 machines a month including some 150 of the 749cc models. At least half of these machines were exported, many of them going to Rudge depots in South Africa, Australia and New Zealand. Velandini in Milan proved to be Rudge's biggest European customer, while other machines were exported to Austria, France, Latvia and Russia, where Rudge's reputation stood the company in good stead for military sales now that the First World War had started. In fact a contract was signed on October 2 1914, to supply the Russian Army with 400 Rudge-Multis. The first batch of this order was built and despatched within a week and the remaining machines were completed and en route to Russia 11 days later.

Meanwhile many of the company's competition riders and staff, once war had broken out, had joined the armed services. Cyril Pullin was now a member of the Royal Naval Air Service, Tommy Greene had joined the motor-transport section of the Army Service Corps, while 'Rowley' Rowlandson had gained a commission in the Army Ordnance Corps. John Pugh's own son Charles, a keen motorcyclist before the war, had also joined the Army.

Rudges at war (1915-1919)

A four-speed gearbox was first used on a Rudge-Whitworth motorcycle in 1915 and not several years later, as subsequent Rudge literature would seem to indicate. It was on January 6 that year that the fourth prototype of the 998cc V-twin was assembled, this time fitted with a proprietary four-speed Jardine gearbox, instead of the Multi gear used on the three previous development machines. It was driven by chain from the engine, while the final drive to the rear wheel was by belt.

This machine subsequently appeared in the 1915 catalogue with the option of either the Jardine gearbox or, as the 1000 Multwin, fitted with the normal Multi gear. Other models featured in this catalogue included the 499cc fixed-drive and clutch models, the TT Rudge-Multi racer, the ordinary Multi and also a sidecar.

With the advent of World War 1, sales of civilian machines had fallen dramatically and only a dozen of the new 998cc chain-cum-belt four-speed solos were actually sold. There was also a disadvantage with these machines in that the engine-shaft-mounted clutch proved particularly tough on gearboxes and gave a lot of trouble. Nevertheless, the basic idea of switching from the Multi gear to a four-speed transmission was valid, since the former could not cope satisfactorily with the power output and torque of a big-twin.

To try to make up for the fall in the civilian market, the firm was actively canvassing for military motorcycle contracts by late 1914. Having successfully completed the order for the Russian Army, Rudge-Whitworth received a second from the Belgian Army for 400 machines, which was

Rudge rider Second-Lieutenant F.M.C. Houghton at the first Brooklands' Combined Services Race meeting, staged on Saturday, August 7 1915.

completed early in 1915. As a result of this the firm had hoped for a large follow-up order from the British Army, but much to John Pugh's chagrin this was placed with the rival Coventry concern, Triumph. The Admiralty did, however, place a small order for Rudge-Whitworth machines to be used by naval personnel in UK shore establishments.

Not only was the home market now shrinking fast, with most active motorcyclists in the armed services, but the firm had now lost some of its most valuable European export markets. True, early in 1915, a small order had come through for a few hundred machines for Italy and quite a few Rudge machines were exported during April 1915 to Copenhagen, including one to the King of Denmark, but this was not able to stop the rot. The same month a Rudge-Multi was presented to the Science Museum in South Kensington, London, where it is still on display. By July 1915, production had dropped to a mere ten a day and these machines were solely for military use. They were largely exported to Russia, Belgium and Italy, with some going to South Africa, Australia and New Zealand.

In 1916 the British War Office was considering a scheme to produce a rationalised, standard, despatch rider's machine constructed from what were considered to be the best components of the leading British motorcycle manufacturers. A number of development machines were produced, for which Rudge-Whitworth supplied wheels and two sets of frames and forks, for experimental purposes. Except for these and one last order for the Belgian Army completed in two batches at the end of 1916 and the other at the beginning of 1917, motorcycle production had now virtually ceased. In fact, on November 3 1916, the Ministry of Munitions had banned the production of motorcycles for civilian use altogether. The Rudge organisation was now working entirely for the war effort, the concern's works in Rea Street producing shells under contract to Wolseley at this time.

A sad happening in 1917 was that John Pugh's son Charles was killed in the Battle of Ypres, something from which John never fully recovered. His personal loss plus the earlier parlous state of company trading did nothing to help his enthusiasm for further developments to help the company when eventually peace was to return. Nevertheless he was to overcome his sense of loss to some extent, enough to come up with some interesting ideas in the post-war period.

By 1917 Rudge wheels were being produced in large quantities for Crossley ambulances, military staff cars and for aircraft landing gear. They were also being supplied to the Italian Army for a most unusual application. At that time the Italians, who were then our allies, were fighting against the Austrians in the Italian Alps. The terrain there made rapid troop movements extremely difficult, so the Italians had organised a series of aerial ropeways upon which four-wheeled trolleys ran. The Rudge concern supplied 7 ft (2.15 m) dia wheels for these. They were fitted with brakes operated by a soldier who sat centrally mounted in a Rudge sidecar. To all intents and purposes it had the appearance of a gigantic perambulator!

The ball-bearing factory was also working to full capacity and had been extended several times to cope with demand. The rest of the Rudge workers had been switched over to the production of ammunition on its existing machine tools. A seventh floor had been added to the main building and the River Sherborne had been decked over to cope with the need for more storage space for finished shells. The Rudge-Whitworth works now bore little physical resemblance to the cycle and motorcycle factory of the pre-1914 years.

With the end of the war in November 1918, Victor Holroyd, now General Manager, made getting back into the production of civilian motorcycles, as soon as possible, his principal target. The firm had not been fully paid for its war work on the understanding that the very much extended ball-bearing factory would revert to the firm's use after the hostilities. He therefore had to find a source of revenue and quickly.

As a stop-gap measure he had managed to get some 25 Rudge-Multis produced for the Italian Army during 1918, but this was clearly a very modest start. With the demobilisation of servicemen with post-war gratuities to spend, the demand for civilian motorcycles was going to be enormous over the next year or so. Many of these young men had learnt to ride motorcycles as part of their duties during the war.

Another problem facing Victor was that many of his pre-war skilled Rudge personnel had been killed during the recent fighting and it was not going to be easy to find suitable skilled replacements in time to provide the necessary upturn in production. He had no easy task on his hands, but he set to with a vengeance. Nevertheless, when the ban on civilian motorcycle production was rescinded and civilian production began again in January 1919, the production rate was a mere two or three machines a day.

An added problem was the lack of British-made magnetos. Importation of the German-made Bosch instruments had ceased at the start of hostilities, so as a stop-gap Holroyd imported a number of American Splitdorf components to get production under way. By July 1919, however, supplies of English CAV magnetos were beginning to displace the imported American product.

At first it was only possible to produce one model — the 499cc Rudge-Multi. It was available in either its Standard or TT Roadster form. By April 1919, sidecars had been re-introduced and, by May, the production rate had been stepped up to something like five to six machines a day, including one or two 749cc 'big singles'. All of these models, though, followed the general specifications laid down in the 1915 catalogue, except for the frame of the TT Roadster which, in contrast to that used on the 1915 model, was of the Pullin-designed 'low frame' type. The Standard Multi and the 749cc machine retained the pre-war 'high frame'. Fortunately bicycle production was very soon back to pre-war levels and this helped tide the firm over the difficult immediate post-war period.

Aware of the difficulties facing so many British motorcycle manufacturers after four years or so of war, the organisers of the pre-war motorcycle shows had staged one in November 1919 to help 'drum up' business. This proved a great success for Rudge-Whitworth, which displayed a total of ten Multi-geared machines including three 998cc V-twins (which had only just been put back into production), two 749cc singles, two 499cc Standard models and three 499cc TT models. The sales resulting from this show were quite good, but there was some worry about the sales of the 499cc models, which with their overhead-inlet, side-exhaust valve arrangement were now being outperformed by their overhead-valve (ohv)-engined, three-speed, all chain-drive rivals. Clearly some serious redesigning was going to be necessary before the problem was solved. Meanwhile, a young engineer and metallurgist had joined the research department of Rudge-Whitworth, who was to have a profound influence on the company's designs and future fortunes. His name was George Hack and he had just been demobilised after a spell of service on the Russian front.

6

A change of gear (1920-1922)

By March 1920 export sales had been re-established and Rudge-Whitworth motorcycles were being sold to countries as diverse as Australia, Belgium, Denmark, France, Greece, India, Italy, Japan, New Zealand and South Africa. The magneto shortage created by the recent war had now disappeared, with the availability of the new ML components and these were now being fitted to the 499cc machines, while 749cc singles and 998cc twins were employing British CAV and American Splitdorf magnetos respectively. Production was now reaching a rate of 20 to 25 machines a day and, by June 1920, had topped 30 a day — the highest rate of motorcycle production ever at the Rudge works. Things were also going well on the bicycle side, especially the sales of lightweight racing bikes, which were proving extremely popular with racing cyclists. In fact all the winners of the 1920 British Amateur Championships were Rudge-Whitworth mounted.

During 1919, John Pugh had initiated the design of a revolutionary new engine with motorcycle speed records in view — a 1700cc air-cooled inline, overhead camshaft (ohc), four-cylinder power unit having 85-mm bores and a stroke in the order of 75 mm. By July 1920 the drawings had been completed. It had finned cylinder barrels, a car-type bolt-on sump and advanced features such as a gear-driven oil pump, shell-type main bearings and slipper-type pistons.

The completed engine was mounted in a frame adapted from the V-twin and fitted with a robust gearbox and chain drive. Apparently the power produced by the engine was exceptionally high for the period and on one occasion whilst on test on a straight stretch of road near Coventry, with the rider flat down over the tank, the sudden acceleration caused the tread to separate from the cover of the rear tyre. There being no rear mudguard, the first the rider knew of this was when his posterior made violent contact with tread fragments as they were being spun off the rear wheel!

The main problem with the engine of this machine was that it was well over the 1000cc maximum capacity limit stipulated by both the ACU and the FICM (as it was then) for racing and record attempts, and at the time these two organisations were unwilling to budge and introduce a higher-capacity class. This led to an acrimonious series of exchanges between John Pugh and those bodies, for he did not see why he should have to redesign the engine to suit what he felt was a purely arbitrary upper capacity limit. As a result he put an unofficial ban on any official works' participation

A 1920 model Rudge-Multi purchased that year and ridden at Brooklands by the author's father, Laurence Hartley.

in racing and records, a ban that was to last for almost two years.

It was a sad story and its final chapter came just before John Pugh's final Christmas at the works many years later when, walking through an almost deserted machine shop, he spotted the old engine. He had had a few rounds of Christmas cheer and was feeling slightly melancholy, and the sight of that power unit stirred such unpleasant memories that he felt the need to exorcise them; so he got two of the remaining factory employees to carry it up to the balcony that ran round the edge of the machine-shop and hurl it down to the machine shop floor beneath. So ended an interesting but ill-fated design as a pile of broken castings.

On the production front, the belt transmission problems with the 998cc twin, first encountered in 1915, still had not been resolved at the beginning of 1920. The Jardine gearbox had proved no satisfactory answer, so John Pugh set to and came up with a gearbox design of his own. To keep things simple, this was given three speeds. An example of its robust construction was the adoption of ¼-in (6.3 mm)-thick selector plates. It was first fitted to a 499cc machine having the standard Multi-type engine, but without the Multi-gear ramp on the crankcase. The adoption of chain final drive simplified the rear-wheel spoking and enabled the outer belt rim to be discarded. Retention of the brake flange, however, and the use of a normal external brake-shoe enabled the introduction of interchangeable front and rear wheels.

The machine was extensively tested throughout the summer of 1920, and proved so successful that a second gearbox was made and fitted to a 998cc twin, on October 27. This again proved successful and a prototype production machine appeared on the Rudge-Whitworth stand at the Olympia Show in November 1920. The other machines on the stand comprised two each of the TT and Standard 499cc Multis, the 749cc big-single Multi and the 998cc Multi twin. A surprise was the option of Lucas electric lighting as an alternative to the traditional standard acetylene system. One each of the 749 and 998cc machines were shown fitted with this new equipment.

The beginning of 1921 saw the Rudge-Whitworth concern enter a boom period. Production was then running at a very healthy 30 machines a day, mostly with the Multi-gear arrangement although the 998cc V-twin fitted with the new gearbox was doing well and there had been requests for a 499cc version. Such a demand had been foreseen by John Pugh, and he and his design assistant, George Hack, had been working on a prototype '500'. On February 23 1921 the decision was made to put this into production and they started to come out of the factory at the end of March. At first, production was to specific customer orders and ran at about five a month. No engine-shaft shock absorber was fitted on the first all-chain drive 499cc Rudges and, as a result, they gave far less comfortable a ride than their belt-driven forerunners.

This problem was solved when, in July 1921, production was started on the firm's 1922 range of machines. On these a spring shock absorber was built into the clutch, which gave some improvement. In August the new gearbox was modified to

O.M. Baldwin on the Rudge he rode in the 1922 Senior TT Race in the Isle of Man.

Bob Dicker with his fast inlet-over-exhaust, but all-chain drive, racing sidecar outfit at Brooklands in 1922.

accept a kick-starter, which was now becoming standard practice on production motorcycles. This was achieved by introducing a solid shaft mounted on two lugs behind the gearbox. The kick-starter and its quadrant gear pivoted on this and engaged a gear on the end of the gearbox mainshaft outside the gearbox end cover. Unfortunately, although this performed satisfactorily, it overstressed the aluminium-alloy gearbox lugs and eventually caused their fracture.

During the period 1920-1921, an enormous number of motorcycle manufacturers had entered the market and over-production on the part of established manufacturers was rife. Many of the smaller manufacturers had folded up by the end of 1921 due to a lack of prospective buyers. Larger motorcycle manufacturers like Rudge-Whitworth were being forced to cut both their prices and production figures. By October 1921, the firm had slashed its numbers of motorcycles produced each day to only 15, of which the numbers fitted with gearboxes and chain drive had risen to 27 per cent. Clearly the belt-driven Multi was on the way out.

At the Olympia Show in November 1921, only two out of the seven Rudge machines on display were in fact belt-driven, a sign of the times. They were the 499cc Standard and TT versions of the Multi. All the others were fitted with three-speed gearboxes, four of them being the newly designed '500', the other the 998cc twin. Sales were now running at an all-time low of five a day.

Clearly some drastic decisions would have to be made to avert a sales catastrophe. John Pugh gritted his teeth and made it known that he was going to reverse his earlier 'no racing' decision. It was decided to enter a team of Rudge-Whitworth machines in the 1922 Senior TT Race in the Isle of Man in an attempt to obtain some good publicity to help boost sales. The problem was, whether to employ the Multi-geared type of machine that had proved so successful in 1914 or go for a team of three-speed racers.

In the end it was decided to enter five of each type, but eventually one of the Multis was replaced by a three-speed machine. Housed in the frames of the Pullin-designed type, to give good all-round handling, their engines were fitted with mechanical Best & Lloyd drip-feed oil pumps mounted on the outsides of their timing covers, the first occasion on which Rudge machines had had positive mechanical lubrication. Auxiliary foot-operated pumps were retained, however, as a back-up. Other improvements included a new cylinder head with a forward-facing induction port and carburettor intake and the fitting of front forks, wheels and brakes from the three-speed models on the four Multis being raced.

Interest in the event was intense at the Rudge-Whitworth factory and great was the disappointment when the first telegram arrived after the race to announce that only O.M. Baldwin, on one of the three-speed models, had successfully completed the course and then only in 14th place. He was the first Rudge rider to complete a TT Race in the Isle of Man on a chain-driven machine and the first to use a three-speed gearbox in that event. All the other nine Rudge team members had retired. After this débâcle, it was decided to disband the Rudge team after the race and no further Rudge-Whitworth team was to appear in the Island for many years to

Bert Mathers astride the 998cc ioe V-twin Rudge on which he and Bob Dicker broke numerous long-distance records in 1922.

come. It resulted in a slump in sales and a resultant drop in production to a miserable seven machines a day.

With the Pullin-type frame now standardised on all Rudge machines, it was difficult to accommodate the tall power unit of the 749cc big single, so it was dropped. This left just the 499 and 998cc machines in production.

Despite the TT failure, private owners were having some success at Brooklands Track in holding up the Rudge marque's flag. One of the most successful of these was R.E. ('Bob') Dicker, who raced a 499cc ioe Rudge with all-chain drive, specially tuned by himself for maximum speed.

Formerly a rider of Norton machines, Bob Dicker's first major outing on the Rudge was at the so called 'Royal Meeting' staged jointly by the Essex Motor Club and the BMCRC on, Saturday, May 20 1922, under the patronage of the Duke of York (later George VI) who had himself entered a machine to be ridden by his chauffeur. The day's racing consisted of a series of three-lap motorcycle races. In the first of these, the Junior (500cc solo) Handicap, Bob Dicker with a 1 min 24 sec start was a hot favourite with the bookmakers. Their faith was well justified for he was extremely fast and looked a likely winner. But so was Captain Maund on his ABC and despite a flying lap at 75.2 mph (121.1 km/h) Bob could only manage a second place.

A week later at the ACU's Midlands Centre Race Meeting at Brooklands a new rider, Bill Lacey, made his début on a 499cc Rudge — he was to become the Brooklands Track's 500cc solo lap-record holder in the next few years.

It was not until Saturday, July 22, at another Essex Motor Club Race Meeting, that Bob Dicker made another appearance at the track on his ultra-rapid Rudge. He rode in two events, gaining second place in both the 500 and 1000cc Solo Handicaps. Then, at the BMCRC's Fifth Monthly Race Meeting of the year on Saturday, August 12, after finishing second in the 500cc three-lap Solo Handicap, he gained his first victory on the Rudge. Quickly fitting a racing sidecar, he went to the starting line for the three-lap 600cc Sidecar Handicap. Getting away well he secured second place early on and then on the last lap passed F.W. Jelpke (348 Weatherell-Blackburne sidecar outfit) to win at 59.15 mph (95.25 km/h).

At the end of August 1922, a spate of record attempts took place at Brooklands with no less than three attempts taking place on the 'Double Twelve' 24-hour record, all under atrocious weather conditions. One of the record contenders was motoring journalist Bert Mathers and his 998cc ioe V-twin Rudge, who on Thursday, August 30, started his attempt, but was forced to retire due to the incessant rain.

It was not until the pre-Olympia record-breaking period that Bert made a reappearance at the track in pursuit of records, this time in partnership with Bob Dicker. The two of them were after the 1000cc Solo Class 'Triple Eight' 24-hour record. On each of three days, October 25, 27 and 28, they took turns riding for eight hours apiece, making a total of 24 hours running with the machine locked away under official seal between times. They set a new record with an average of 61.72 mph (99.39 km/h) and covered some 1481 miles 459 yards (2385.3 km).

On Saturday, November 11, three motorcycle races were featured in the Remembrance Day Brooklands Race Meeting. The last of these, the Weybridge 90 mph Handicap, was fast. Rumours that Bob Dicker's 499cc ioe Rudge had been lapping at nearly 80 mph (129 km/h) in practice, made it a hot favourite with the Brooklands' bookies. In the race itself the best he could manage was second place behind O.M. Baldwin's big 992cc Matchless-MAG.

As always, the end of the racing season at Brooklands attracted a lot of record attempts and Bob Dicker and Bert Mathers teamed up again to have a crack, this time, at Bert Le Vack's 500-mile 1000cc solo record, riding their 998cc V-twin Rudge machine. Starting out in the early hours of the last day available for record attempts, before the track closed for its customary winter repairs, Saturday, November 25, each took shifts in the saddle, lapping consistently at around 76 mph (122 km/h).

At six hours the Class E (1000cc solo) record for that period fell to them at an average speed of 75.02 mph (120.8 km/h). They went on to set new 500- and 600-mile records as well, at 74.96 and 71.27 mph (120.7 and 114.8 km/h).

These long-distance records demonstrated the V-twin's stamina, but its low peak engine speed of 3600 rev/min really demanded a four-speed gearbox, something that was to be next on John Pugh's design agenda.

At the Olympia Show three basic types of 1923 model were put on display, the 499cc Multi, the 499cc three-speed machine, with the option of ML or BTH magnetos, and the 998cc three-speed big-twin. Something that particularly interested the Duke of York when he visited the Rudge-Whitworth stand at the show, was a miniature motorcycle and sidecar that was only 3 inches (76 mm) high. It was later presented to Queen Mary, his mother.

7
Four speeds and four valves (1923-1924)

The first prototype 499cc machine to be fitted with one of the new four-speed gearboxes was hastily assembled on January 16 1923 and sent for display at the Glasgow Exhibition. With sales falling it was essential to get things moving, for it was considered to be too long a wait until July, when the 1924 range of models would be announced, before the new gearbox went into production. It was not evolved from the three-speed design, but was a quite new concept. Compared with previous gearbox designs it was 'upside down' in that the layshaft and kickstarter spindle were beneath the mainshaft. Its securing bolts were at the top of the gearbox casing. The selector quadrant was located at the rear of the box, while the operating arm was mounted on the nearside behind the sprocket and connected by a rod to the projecting end of a cross-over shaft on the nearside of the petrol tank. On the offside of the tank, the projection of this shaft was fitted with a gearchange lever.

Tests of the machine on its return from Glasgow after the exhibition, using a 4:1 top gear, proved successful, so, on February 9, a 998cc twin was also fitted with a four-speed gearbox and put on test, this time with a 3.5:1 top gear.

With the lack of sporting successes at this stage of its career, the Rudge-Whitworth company was attempting to gain interim publicity by interesting well-known people of the day in riding the firm's machines. Rudge motorcycle catalogues of the period were studded with famous actors and actresses sitting on their Rudge-Whitworth machines. It therefore came as an additional boost to the firm's campaign when it became known that the third of the new four-speed 998cc twins had been supplied to the famous stage comedian Max Miller. At the end of February 1923, the four-speed models were in full production and the three-speed types virtually phased out.

Bob Dicker and Bert Mathers continued to uphold the firm's prestige at Brooklands and at the First 1923 BMCRC Members' Race Meeting on Saturday, April 7, Bob gained a third place in the 500cc Solo Handicap despite a heavy time penalty at the start. Two months later, on Wednesday, June 6, Bert Mather and Bob Dicker, with a sidecar attached to their 998cc ioe chain-drive Rudge, set out again after records. The principal objective in view was the Class G (1000cc Sidecar) 500-mile record set up the previous September by H.H. Beach (588cc Norton sidecar) at an average speed of 52.46 mph (84.48 km/h). This was well and truly broken with a speed of 58.88 mph (94.81 km/h). Other Beach-held records to fall to the Rudge duo that day included the seven- and eight-hour speeds, taken at 59.23 and 58.49 mph (95.38 and 94.19 km/h) respectively.

Not content with this, Mathers and Dicker brought their outfit out again on the Friday morning to attack the Class G six-hour record held by Bert Le Vack (998cc Indian sidecar) at 60.57 mph (97.54 km/h). Mathers took the first shift in the saddle, lapping at a regular 68 mph (109.5 km/h) until, after 18 laps, the high-tension lead on the rear cylinder came adrift. Luckily this happened close to the pits, enabling not only a quick replacement, but also tank replenishment with fuel and oil, and he was soon away again. Dicker took over at 50 laps and the lap speed was raised to around the 69 to 70 mph (111 to 113 km/h) mark, to neutralise the effect of the stop. An hour later the two men changed over again, though by this time it had begun to rain heavily, reducing lap speeds appreciably. At 12:15 pm the Class G three-hour record fell to them at an average speed of 66.23 mph (106.7 km/h) then, a few minutes later, the 200-mile record of 66.39 mph (106.9 km/h).

On the 91st lap the back tyre burst, as Mathers was travelling on the Byfleet Banking at the Southern end of the track and he coasted into the pits on the rim. However, a quick change of the back wheel and he was away again with little time lost. But the rain was now really starting to pelt down.

Another four Class G records fell in close succession despite the conditions — the four and five hours and the 300 and 400 miles — all at around the 64-mph (103-km/h) mark; then at 3:15 pm the target was reached and the six-hour record fell after a distance of 383 miles 989 yards (617.65 km) had been covered, representing an average speed of 63.92 mph (102.9 km/h). It was decided to keep going until something 'broke', despite the appalling weather conditions; however, the onset of a real gale made further running dangerous, so matters were brought to a close at 151 laps, after a highly successful day of record breaking.

Nineteen-twenty-three was a disastrous year for the motorcycle makers, there being too few customers and too many machines on the market after the previous two years' excessive production. Many small manufacturers had just 'gone bust' and disappeared from the scene, while the more stable, larger ones were engaged in a price-cutting war to stay alive. Rudge-Whitworth was not exempt from this trend either and to illustrate the state of affairs, the price of the 499cc three-speed model dropped from £94 in 1922 to £80 in 1923, with the four-speed machine of the same capacity, when introduced, costing a mere £5 extra. Similarly the big twin had dropped from £113 in 1922 to £100 a year later in its three-speed version. Strangely the four-speed version of the big-twin cost only another £2 10s extra. The overall machine production for the year 1923 was a disastrously low 1400.

Clearly, if the firm was to survive this crisis and compete successfully with other makes on the market, it would have to have that something extra to offer the potential buyer. Motorcyclists by and large were (and still are) a sporting crowd and demanded good performance from their machines and were now in a position to call the tune in that respect. The old overhead-inlet, side-exhaust Rudge engine was just not competitive enough, so John Pugh turned his thoughts to redesigning the engine with performance in mind.

Throughout 1922 one of the most successful new machines in motorcycle racing had been Major Frank Halford's 499cc four-valve, single-cylinder Triumph. This had an engine designed by Sir Harry Ricardo and was widely known as the 'Ricardo-Triumph'. Initially air cooled but later water cooled, it had four valves arranged in 'pent' or 'V' configuration, with two parallel exhaust valves at the front and two parallel inlet valves at the rear. With pushrod operation, the engine was quite powerful for the day and scored a number of Brooklands successes in 1922 and 1923. Unfortunately the cycle parts did not afford commensurately good machine-handling so that there were no corresponding road race successes.

John Pugh was attracted by the valve layout. It would remove the risk of possible cylinder-head distortion that he feared would occur if he employed a simple two-valve ohv layout with valves large enough to provide the necessary performance he was seeking. The four-valve head also gave other advantages. The overall inlet-valve area and, hence, 'breathing efficiency' would be increased, the lighter valves would enable higher engine speeds to be reached and, compared with the former ioe engine, a much higher compression ratio could be employed. All three factors in combination would result in higher power and an engine of this type housed in the excellently handling Rudge frame should provide a formidable sporting road machine and should sell well.

With these thoughts in mind, blueprints for a new four-valve Rudge engine were prepared and the first experimental version assembled on July 27 1923. It had the traditional Rudge ioe engine's bore and stroke of 85×88 mm, giving a capacity of 499cc.

Each of the four valves had a 1-in (25-mm) diameter head and rather thick, 0.37-in (9.39-mm) diameter stems. They were operated by pushrods via overhead rockers from a camshaft with a single-lobe cam and a cam follower on each side. The camshaft and big-end assembly were almost identical to that of the earlier ioe engine, while the magneto was mounted on the front of the crankcase and driven via a train of gears. The first gear in the train, however, also drove the Best & Lloyd constant-loss oil pump which had so proved its worth in the Isle of Man Races.

The performance of the engine exceeded all expectations and on test it produced some 15 bhp, which was 50 per cent over the best obtainable with the production ioe engines. This led to unforeseen problems with the cycle parts which although able to cope with the power output of the former ioe engine were unduly stressed using the new engine. To cope with the greater efficiency of the new engine without involving a major redesign of the whole machine, the engine was redesigned as a '350' in September 1923, with a bore and stroke of 70×90 mm, and the cycle parts were strengthened.

The relatively complex linkage system for the gearchange mechanism was now placed on the offside of the machine, pivoting on the frame beneath the saddle and doing away with the earlier cross-over shaft. The horizontal tube in the tank, in which this shaft operated, was retained though, and a spring mounted inside it to give the gearchange lever, mounted on the side of the petrol tank, positive location within the indentations of the gate. Then in October, an entirely new frame was produced, having an additional torque arm running from the rear spindle mounting to the bottom of the crankcase. The steering-head cups were redesigned and their ball bearings increased in size from $3/16$ to $1/4$ in (4.8 to 6.4 mm). This arrangement was retained on all Rudge-Whitworth ohv machines until production of them stopped in 1939. A further change was the brazing of the handlebars into the steering lug, which was held to the top of the front fork spring and the rear of the top shackle. Till then the traditional cycle practice of securing the handlebars by a bolt passing through their centre and into the steering stem had been used.

The first two prototype machines were assembled on October 9 1923, so as to be ready for display at the Olympia Motorcycle Show the following month. A third prototype, assembled on October 15, was kept back at the factory to enable testing to continue. Other redesign features brought in in October included a front mudguard with a width of 5 in (127 mm) and having a flattened 'U' shape, henceforth one of the Rudge-Whitworth hallmarks. Initially retaining the hinged rear flap for wheel removal, the rear mudguard's shape in cross section was also modified to a flattened 'U' shape.

So much work and time had gone into developing the new 350 that the two 499cc four-valve machines on display had little modification and retained the earlier four-speed gearbox with cross-over gear operation and unmodified frame. Although for continuity, a solitary 499cc Rudge-Multi was shown at the exhibition, the very last Multi was assembled at the Rudge-Whitworth works on October 31 1923. It is interesting to note in this context that the 1924 catalogues of the Italian Moto Borgo company showed photographs of the Moto Borgo Tipo Turismo model which appears to have been a 499cc Rudge-Multi in all but name, complete with ioe engine and Multi gear. The tank, true, had the emblem 'Borgo Cambio Graduale', but, according to Erwin Tragasch in his book *The Illustrated Encyclopedia of Motorcycles,* Carlo and Alberto Borgo had a measure of co-operation with Rudge-Whitworth at this time and it could be that this involved an agreement to enable them to produce the Multi under their own name under licence in Italy even though production had stopped in the UK.

Rudge-Whitworth also had on display at Olympia, presumably also for continuity, two four-speed 998cc V-twins, one with electric lighting. Here again, production had nearly terminated as the last of the V-twins was assembled on November 2 1923.

After the show, one prototype four-valve machine was sent for display at the Brussels Motorcycle Show and the other to the firm's Italian agents, Velandini, in time for display at the Italian Motorcycle Show in Milan.

During the last two months of 1923, the factory was retooled and reorganised to take account of the new models in the pipe line. A Mark IV prototype '350' four valver was assembled on Boxing Day, and production of current models restarted on New Year's Day 1924 at the rate of 35 machines a week — the machines still being fitted with the older 499cc ioe-type engines and the older-type gearboxes and frames. Two of these models were shown at the Scottish Exhibition the same month, before their production finally ceased and gave way to the new '350' four-valve machines on January 18 1924, when the first three new production models were assembled.

The firm had thrown all its 'eggs' into one basket in designing and producing the new '350' four-valve machine. All its development resources had been locked into getting the new machine into production as soon as possible, but there was still a very successful sidecar being produced and there now existed an obvious need, with the demise of the ioe '500', for a new '500' model to power it. With this in mind John Pugh and George Hack applied their minds during February and March 1924 to designing such a machine to supplement the new '350'. By the end of April, the new '500' was ready for production. Its frame was similar to that of the '350' except that, because the nearside gear-quadrant-controlled gearbox was employed, it was necessary to use a cross-over shaft just beneath the seat pillar to retain an offside, tank-mounted, gate-change lever.

The new '500' engine was slightly different to the '350' in that it had a camshaft with both an inlet and an exhaust lobe attached. This enabled the push-rods to be closer together both at the top and the bottom of the cylinder, whereas the arrangement on the '350' caused them to be more widely spaced at the top end than at the bottom.

Turning the gearbox arrangement upside down on the '500' raised the clutch centre higher than that of the crankshaft, so that the pressed-steel, primary chain cover had to be tilted up towards the rear. The new clutch, with its large central spring, projected through a hole in the primary cover and so had to have its own rotating protective cover.

As if all these radical design changes were not enough, a train of new design improvements was set in motion in July 1924, ready for incorporation in the 1925 models due to be shown at the Olympia Motorcycle Show in the autumn. They included a complete redesign of the gearbox, the modification of the rear carrier and mudguard, and the fitting of footboards as standard on the '500' with the option of footrests as an alternative, if specified. The kick-starter and layshaft were now to be at the top of the gearbox, giving a lower mainshaft and clutch while, instead of just a flap on the rear mudguard, the whole unit behind the saddle could now be pivoted forward to assist back wheel removal. Then, in

October 1924, came the *pièce de résistance* — the introduction of coupled front and rear brakes operated by a single foot pedal.

Both the '350' and the '500' handled well, but with only low compression ratios they only produced maximum power outputs of 10 and 15 bhp respectively. Nevertheless, it was not long before the exploits of Rudge riders on these two models started to appear in the Rudge promotional material. Quite a few were exported and R.E. Canteloup, a New Zealander, was soon being headlined in Rudge publicity material due to his taking his '350' model, to the accompaniment of much 'footing' up the soft volcanic ash slopes of 1200-ft (369-m) Mount Eden, an extinct volcano near Auckland.

The changes in design were clearly paying dividends since demand for the new Rudge-Whitworth machines was rising rapidly and, during 1924, production had to be increased to 3000 machines to cope with this — more than double the previous year's output. The sidecars were also selling well and just before the Olympia Show the standard three models produced by the firm were joined by a fourth — a long-nosed design called the 'Boat Shape', which was as long as the machine itself. Prices were now remarkably low, bearing in mind the technical innovations the machines incorporated, with the '350' costing £42 and the '500' only £4 more. For the expenditure of another £4, electric lighting could also be fitted as an optional extra.

In October 1924 a completely new prototype '350' machine was designed, with a view to it being the eventual production model to incorporate the new coupled braking system. It had several novel features. Firstly, its crankcase was split in conventional vertical style, but only to just below the main bearings, where a conventional car-type sump was fitted containing the oil pump. An extremely modern design feature was the adoption of a positive oil feed to the roller-bearing big end via the crankpin. Another departure from the, till then, normal Rudge practice was the adoption of a 71 mm bore and an 88 mm stroke, enabling the use of identical crankshafts on both the '350' and '500' models, if ever the former were to enter production (which it did not).

The prototype that superseded this '350' was more conventional, having cylinder dimensions of 70×90 mm. It was identical to the 1924 production '350' except for the fitting of a two-valve cylinder head and the placing of the rocker-return springs on top of the rockers rather than between their undersides. This was intended to reduce the chance potential loss of a pushrod due to valve bounce caused by the previous arrangement, which was intended to reduce the pressure on the cams. Unfortunately the new arrangement, conversely, increased cam and follower wear. There was no real solution to this dilemma, so the spring was eliminated completely from the design on all models at the end of 1924.

The final (and third) prototype engine was simply a 1924 four-valve production unit with a redesigned crankpin and big-end assembly, and this eventually went into production at the end of 1924 for the 1925 season.

8

Back to racing (1925-1927)

A rethink on the benefits of competition was clearly in the air at the Rudge-Whitworth works towards the end of 1924. Symptomatic of this was the long-distance record-breaking efforts on a 1924 model '350' four-valve Rudge at Montlhéry by the then famous duo of Colonel Neil Stewart and his wife Gwenda. On March 30 1925, starting at 10:05 am and taking two-hour spells each in the saddle, they broke a string of long-distance world records up to 24 hours duration in the 350, 500 and 750cc solo classes. Except for a foot-operated oil pump and the fitting of an extra-large fuel tank to reduce the intervals between refuelling stops, the machine was virtually standard and averaged 54.21 mph (87.29 km/h) for the full 24 hours of running time, covering a distance of 1301.14 miles (2095.23 km) — a commendable feat of endurance both for the riders and the machine.

In the first few months of 1925, the new production machines used cycle parts similar to those of the 1924 range. In March 1925, experiments were carried out with a bolt-up fork assembly, the flexibility of which merely confirmed the efficacy of the standard Rudge-Whitworth rigid fork legs and shackles. The next step occurred in April that year when the torque stays, which had proved unnecessary, were discarded. This allowed the chain stays to be lowered and bolted to the engine plates immediately behind the rear of the gearbox. This in turn enabled the gearbox to be more easily removed. Another modification was the elimination of the seat-tube-located gearchange pivot, the tank-side-mounted gearchange lever now being connected directly to the gearbox operating lever.

Another, but slightly different development was a canoe-shaped sidecar having the amazing overall length of 16 ft (4.9 m)! The idea, apparently, was that canoeing enthusiasts could, by undoing quick-release clamps, remove it from the chassis and use it in the role envisaged by its name. Over the years, the Rudge-Whitworth designers, possibly John Pugh himself, certainly seemed to have had rather a fixation on boating in some respects!

G.E. ('Ernie') Nott, then a works tester and a very tough man indeed, took a Rudge outfit fitted with one of these monstrosities home to a bronze medal in the Birmingham Club's Victory Trial that year on March 7. With an enormous turning circle of some 20 ft (6.15 m), however, it required the services of two other Rudge team members to man-handle the outfit sufficiently for him to negotiate the more difficult parts of the course.

In July 1925, the Rudge-Whitworth motorcycle range for the coming year was announced. The decision had been made earlier to drop the non-competitive 350cc model, so for 1926 only '500s' were to be produced. This specialisation enabled more attention to be paid to the two basic but conflicting requirements of motorcyclists by developing two quite different versions of the model. The first was an all-weather machine fitted with footboards and legshields, the second was one with footrests as a sporting optional extra to footboards. Also, the four-speed gearbox had been redesigned to take the quieter parallel-toothed pinions, whose manufacture had been facilitated by gear cutting developments which had taken place during the year.

The two versions of the 499cc machine were put on display at the Olympia Motorcycle Show that year, together with a full range of sidecars. The chassis of the latter had been the same since their first introduction but, for 1926, John Pugh and George Hack had come up with the novel improvement of replacing the formerly rigid tubes with ones of spring steel, so that when the chassis bent under cornering stresses it returned to its original shape when the outfit straightened up again. It proved a little unnerving for the sidecar passenger at first, but nevertheless improved the handling of an outfit enormously.

By the end of 1925, sales of Rudge-Whitworth machines were quite definitely on the up-turn and, to meet the increased demand, production had reached a grand total of 7000 machines by December 31. Against this background and the reasonable success of its machines in trials and similar competitions, the firm decided the time was ripe to make a considered approach to re-entering road racing during 1926. Not wishing to repeat the TT fiasco of three years before, it decided to do so at first by giving support to private entrants. With this in mind, in December 1925, a number of racing components were designed, including: special handlebars, a forged Elektron magnesium-alloy racing piston (to replace the cast-iron one used previously) and a novel method of damping the

This machine has been dated as a 1924 model by the National Motor Museum, but it does not have the front-mounted silencers of that year and is almost certainly of 1926-27 vintage.

steering laterally against speed wobbles. The latter consisted of a metal band incorporated within the steering head, between the two steering cups. A friction material covered the inner surface, through which passed the fork stem. A bolt was attached to this band and passed through a hole in the offside of the steering head, in which it was held by means of a wing-nut just below the upper steering cup. Tightening the wing-nut tightened the band round the fork stem to give highly effective friction damping. Since the wedge-shaped petrol tanks then used had no bulbous front, these nuts were easily accessible. The introduction of the bulbous-nosed saddle tank put paid to the use of this device after 1927.

In May 1926 the Birmingham firm of Frank Whitworth Ltd, with works support from Rudge-Whitworth, entered two Rudge-Whitworth machines in the Isle of Man Senior TT Race, scheduled for Friday, June 18, the riders nominated being Frank Whitworth junior and Stan Ollerhead, with Ernie Nott as the reserve man. These machines were essentially standard models modified for the race.

The usual overhead four-valve, pent combusion-chamber, pushrod-operated engine, with centrally disposed sparking plug, was used of course; but the compression ratio had been raised to 6.3:1 using a dome-top aluminium alloy piston of new design in which two rings only were used, with a chamfered portion below the bottom ring groove and communicating holes to the interior to provide for its lubrication.

The modified Senspray carburettor used was equipped with a twist-grip control in which the wire issued from the hole in the handlebar normally used for the front brake cable. Since this layout prevented the usual inverted lever being used, the hand-operated front brake control lever became a duplicate of the clutch lever on the offside handlebar. On the nearside 'bar was mounted the ignition advance lever, the clutch lever and the inverted exhaust-valve lifter.

Close ratios had been fitted to the four-speed gearbox and the clutch had two driving plates instead of just one, to give easier disengagement. Also employed was the Rudge-Whitworth coupled-brake system, operated by double pedal — one on each side of the machine.

Although the petrol tank was of the customary

The 1926 '500' Sports Rudge with four, parallel overhead valves (parallel inlets and parallel exhausts).

Rudge-Whitworth shape, it was considerably wider than standard to give a capacity of 2½ gallons (11.37 litres). The oil tank was carried on the saddle tube, a Best & Lloyd mechanically-operated oil pump being driven from the camshaft in addition to a heel-operated plunger-type pump. Other details included the use of an ML magneto, a gate gear-change on the petrol tank and wired-on 27 × 2¾-in (686 × 70 mm) Dunlop tyres.

During the first week's practising all three Rudge riders rode well and were noted as being amongst the fastest riders. On Saturday June 5, Frank Whitworth made the fastest Senior practice lap in 35 min 25 sec, while over the Sulby Straight he clocked 79 mph (127 km/h). Ollerhead returned a speed of 80 mph (129 km/h), some 5 mph (8 km/h) slower than the fastest '500' though. On the last day of practice, Ollerhead got on to the leader-board in second Senior spot behind Varzi on a Sunbeam. The weather conditions were foul, though, so this was no true indicator of performance. The general view of the Rudge machines at the end of the practice period was that they seemed reliable but were lacking in speed. So it proved in the race as well, with Ollerhead coming home in 13th place and Frank Whitworth junior in 15th spot. Clearly some more speed would have to be found before the next year's event.

Meanwhile things were going a little better for the marque on the trials' front with some four gold and seven silver medals accruing to Rudge riders out of the 17 that rode in the London-to-Land's End Trial earlier in the year.

A month after the Isle of Man TT, in July 1926, the company introduced its range of 499cc models for the coming year. As introduced, they had large-diameter clutches with four friction discs and the secondary clutch plate riveted to the chain wheel. Only a few machines were in fact fitted with these clutches, since very shortly afterwards a more substantial type was designed, of large diameter but much slimmer, which allowed a reduction to be made in the height of the primary chain cover. This clutch had four friction discs, two riveted to the chain wheel and two others to a secondary plate having tongues slotted into the chain wheel. This gave very smooth operation and it remained a standard Rudge fitting for many years, only being redesigned when it proved incapable of coping with later engines' power outputs.

In September 1926, the crankcase and its timing cover were also redesigned and the Best & Lloyd external oil pump replaced with an internal oil pump of Rudge design, housed inside the timing cover. Once again, this was but a flow-metering device like the pump it replaced. Mounted vertically it was driven off a worm gear on the end of the intermediate pinion of the magneto drive train. An oil 'tell-tale' plunger was mounted on the timing cover in addition; a feature incorporated in all succeeding models as well.

These various features plus a new type of gearbox were all incorporated into the new 1927 models shown on the Rudge-Whitworth stand at the Olympia Motorcycle Show in November 1926. Superficially similar to its predecessor, this gearbox had many detailed modifications, some of which proved necessary as teething troubles were revealed in subsequent tests. In particular, the second- and third-speed gear selectors gave trouble. These ran in grooves which had teeth on each of their sides. Since these grooves were off-centre, the teeth on one side were rather short and with bad gear changing could break off causing a lot of expensive damage to the internals of the gearbox. This design defect was soon remedied after a few expensive gearbox 'blow ups' had occurred.

The finalised gearbox, whose smooth positive operation was the result of the extensive use internally of needle roller bearings, continued in production until 1939.

Lastly, attention to cylinder-head design brought about the much needed improvement in power output shown up by the TT races. Part of this redesign involved the use of longer valves with slimmer ⅜-in (9.3-mm) diameter stems. However, improvements in one direction frequently show up deficiencies in another and in this case it was with regard to braking. The brakes used on the 1926 machines consisted of two external shoes acting upon the outside of a dummy belt rim. They were hopelessly inefficient for stopping an 80-mph (129-km/h) '500'. With the prospect of active participation in road racing in 1927, some urgent design action was called for in this department.

In April 1927, the first drum brake designed by John Pugh and George Hack appeared, and amazingly advanced it was too. First of all, the drum itself had a diameter of 8 in (203 mm) and contained six shoes — yes six! — each ½ in (12.7 mm) wide. Each shoe, therefore, covered 60° of arc within the drum and was held at each end by a spring. They were actuated by means of a cam plate behind the shoes and concentric with the hub centre. As the lever was lifted, the cam plate rotated forcing the brake-shoes against the drum. This arrangement, however, was somewhat complicated and was eventually redesigned to accept only two shoes.

It was around this time that the company lost one of its leading lights — Bernard Vernon Pugh — who looked after its financial affairs. His death and the forthcoming economic depression were at least two of the factors that were to bring the company to insolvency during the next six years.

Rudge-Whitworth decided to enter a team of machines in the 1928 Senior TT Race in the Isle of Man and George Hack was appointed its Racing Manager. Frank Longman and Cecil 'Count' Ashby were engaged as official works riders together with one of the firm's own employees, Ernie Nott, who had been a reserve rider the previous year.

The company's thinking on what should constitute a high-speed machine received an early airing with the announcement of a new 100-mph (161-km/h) Rudge-Whitworth racer at the end of March 1927. Intended for limited production for despatch to special order, largely overseas, it had a gearbox and cycle parts very similar to the previous year's TT racers. Most of the modifications from standard had taken place within the engine. The compression ratio had been raised to 7.8:1, while inlet ports in the four-valve engine had been enlarged and polished to increase engine 'breathing'. The exhaust ports, which had also been enlarged, were now disposed radially instead of being parallel. The sparking plug, although located in the centre of the combustion chamber bowl, was nearer to the inlet valves than the exhausts, to avoid cracks developing due to the exhaust port walls being too thin at the hottest part of the cylinder head.

The flywheels had been machined and polished to help reduce oil drag. The big end had the standard Rudge-Whitworth ten long rollers, each of ³⁄₁₆ in (4.76 mm) diameter, housed in a phosphor-bronze cage.

The valves were of standard size, but with slightly hollowed heads to lighten them, without affecting combustion chamber shape and volume unduly. Air-fuel mixture was supplied by an Amac twin-float chamber carburettor and an ML barrel-type magneto provided the 'sparks'. On test running on alcohol fuel, the engine produced a peak output of 28 to 28 bhp at some 5200 rpm. Ernie Nott had suggested that the new engine was fast enough to justify an attempt by him to be the first to achieve 100 miles (161 km) in the hour on a '500', but this was rejected by George Hack on the grounds that he

Stan Glanfield after having successfully completed his epic record-breaking trans-world motorcycle trip via India and Australia.

had insufficient track experience to make the attempt.

On the works TT machines, because of the length of the Isle of Man race, a larger petrol tank than standard was fitted. Unfortunately the frame tubes restricted the maximum capacity possible using the wedge-type tank, so it was replaced by a saddle tank with similar square lines, which bore at least a passing resemblance to the standard production machine's component. The new internally-expanding hub brake was fitted, but most of the important design changes had occurred within the engine.

The TT machines were fitted with the new cylinder head developed for the export racer and for the TT a compression ratio of 6.5:1 was selected. Also, like the export racer, the engine was fitted with a flange-fitting Amac carburettor, because of the inadequacies of the Senspray instrument with the latest engine designs.

Unlike the 1926 TT machines, the new ones were very fast, but had less reliability. The result was three retirements in the race. Persevering, though, the Rudge team went on to the Continent and gained a second place in the Belgian Grand Prix with Ernie Nott in the saddle and a second place in the Ulster Grand Prix with Frank Longman piloting, while 'Count' Ashby gained a third spot in the Grand Prix d'Europe, run that year at the Nürburgring in Germany. Clearly, by the end of the 1927 racing season, some degree of reliability had also been built into these 100-mph (161-km/h) machines.

While all this work had been going on in the racing area, the sales side had been undergoing something of a revolution. Rudge-Whitworth had published a new 124-page book called *The Rudge Book of the Road* which, amongst other things of a more general motorcycling interest, covered various aspect of owning and riding a Rudge-

Whitworth machine, all written in a humorous and entertaining vein likely to appeal to the average rider. At the same time, at the end of the year, John Pugh managed to persuade Graham Walker, then with the Sunbeam concern, to join his Coventry team as Sales Manager. Meanwhile, a dramatic re-design of the whole Rudge motorcycle range was underway.

The standard range of Rudge-Whitworth machines for 1927 comprised a 499cc single in three basic forms, Standard, Special and Sports. The differences between them were essentially cosmetic to facilitate production. The Special was a de luxe version of the Standard model which, with electric lighting, cost £50. The Special, costing £5 more, had an adjustable André friction damper on its front forks to damp up and down movement, while the Sports machine, which had nickel-plated exhaust pipes and aluminium-alloy silencers with nickel-plated tail pipes, was fitted with a specially tuned 28 bhp engine. The Standard model had an expansion box with a black tail pipe, mounted just in front of the crankcase.

An unusual item in the 1927 catalogue was the Rudge caravan, available for 130 guineas (£136.50) complete with motorcycle and semi-sports sidecar. This caravan was 7 ft 3 in (2.23 m) long, 4 ft 10 in (1.49m) wide and 4 ft 7 in (1.41 m) high. It had two fitted divan beds, a table and various storage lockers for food, clothing etc. Being constructed of wood, it had no internal arrangements for cooking on account of the fire risk. In fact it was recommended that washing facilities should be added by the owner, since they were also absent, while cooking should be carried out outside the vehicle, presumably on a Primus stove or something similar. There were also commercial versions, some of which are reported to have still been in use as late as 1944 by the Coventry Co-op for delivering milk.

Two of the most remarkable performances by Rudge riders in 1927 were those of Flight Lieutenant S.W. Sparkes and Stan Glanfield, a well-known motorcycle dealer. Starting out in company on their sidecar combinations they proposed to motorcycle around the world. The plan took them across Europe to Turkey, through Iraq to India, where they disagreed about the route to take from there. Stan decided to take the longer trip passing through Malaya, Australia and North America. From India he went to Ceylon and then took ship with his outfit to Australia, then on to San Francisco, crossing the USA to New York and sailing home to England to a reunion with the good Flight Lieutenant some 120 days after they had started out on their marathon trip. The epic journey was full of incident. Stan Glanfield, in particular, encountered seemingly endless natural hazards. Thus, in the desert near Basra he found himself caught up in a quagmire of oil-rich sand and had to manhandle his outfit through it for more than 20 miles (32 km) before getting out of it — he had been on the site of what was to become one of the richest oil fields in the world. In Australia, miles out in the 'bush', his sidecar wheel hit a tree stump buried under dust and sheared off at the spindle. As a result his outfit turned turtle several times and one of the handlebars stabbed him in the leg in the process. After some rudimentary first aid he set out to find his missing wheel, eventually locating it some 440 yds (400 m) back along the trail. Fortunately he had a spare spindle and was able to motor gingerly to the nearest town for some very necessary medical aid. On completion of his trip, he discovered that he had completed a grand total of some 11,650 miles (18,760 km) over road and track, mainly the latter.

9

Rewards from racing (1928)

The works' racing team now consisted of 'Count' Ashby, Jack Amott, Frank Longman, Ernie Nott, Graham Walker and a young Irishman from Dublin University named H.G. Tyrell Smith, who had finished 13th in the 1927 Senior TT Race in the Isle of Man on a Triumph.

The very much improved performance of the Rudge-Whitworth works' racing machines became apparent quite early on in the season, in May, when Ernie Nott cruised home to a relatively easy victory in the Athy 75-mile (120.8-km) Road Race's 500cc class, at an average speed of 62 mph (99.8 km/h), with young Tyrell Smith putting up the fastest lap at 65 mph (104.7 km/h). The result also suggested that some reliability had also been gained without loss of speed.

The Senior TT Race in the Isle of Man, however, proved extremely disappointing. Ernie Nott had again demonstrated the speed of the new racing models by coming top of the leader board in one of the practice periods, with a very respectable lap time for the day of 33 min 33 sec, some 14 sec faster than his nearest rival Charlie Dodson on the Sunbeam. In the race, however, the event became a catalogue of disasters for the Rudge-Whitworth team. Ashby was a non-starter, while both Graham Walker and Ernie Nott retired, Graham while in the lead. Tyrell Smith had slightly better luck, finishing fourth, while the fourth team member, Jack Amott, came home in 11th spot. Graham Walker was particularly upset as he fully expected to win the event.

Now, as has been mentioned, the Rudge engines then used for road racing were, basically, modified standard-type machines, lubricated by a constant-loss system in which the oil was fed from an externally mounted Best & Lloyd pump modified to include a metering jet to control the flow of lubricant. Now, based on his racing experience with Sunbeams, Graham Walker had asked for his machine to be supplied to him fitted with an auxiliary footpump so as to ensure that at all times the big end of his engine got an adequate supply of oil; but this the Rudge works 'deemed unnecessary' and turned down his request. The real truth was that such a device had been fitted as standard on the earlier Rudge machines and dropped some four years before. To go back to using a foot-operated pump was considered retrogressive and bad publicity to boot — so there it was. As it turned out, this omission was to prove Graham's Achilles' heel in the Island Race itself.

On the last day of practice, Graham Walker's machine had been using large quantities of oil and had a very smoky exhaust. No one could quite pinpoint the reason for this, so the Rudge Team Manager and Development Engineer, George Hack, readjusted the pump setting to reduce the oil supply and try to cure the problem. It seemed to work, for there were no more smoky fumes being emitted from the exhaust afterwards.

The 1928 Senior TT was quite one of the wettest on record. All thoughts of record-breaking speeds had vanished and riders were retiring in droves throughout the day. Jimmy Simpson led on the first lap but retired on the third letting Charlie Dodson into the lead on the works' Sunbeam. He hung on to his lead in the fourth and fifth laps, hotly pursued by Graham Walker. Then, on Lap 6, Charlie slowed and Graham took the lead. Earlier he had come in to refuel and was more than a little perturbed to find that his engine had used next to no oil. Since there was no external pump adjustment and the foot-operated pump he had wanted had not been fitted, he decided just to keep going but to carefully 'nurse' his engine. As he passed the pits on his last lap he was mistakenly given the signal: 'You have only 3 seconds' lead, go like hell!' In fact, he had a good 3 minutes' lead over Charlie Dodson.

With many misgivings, he obeyed the instruction to the letter. His qualms were justified, for on the last lap at the Gooseneck, the big end gave all the signs of incipient seizure, so he throttled back hoping to make the finish in first spot. Then with only 9 miles (14.5 km) to go, it happened and the big end went up solid putting him out of the race and handing victory to Charlie Dodson on a plate. It was a bitter pill for Graham to swallow since he had been a member of the Sunbeam racing team the previous year. Incidentally, the big end that had caused all the trouble was made up of alternate bronze and steel needle rollers.

Immediately after the TT, undeterred by the problems it had raised, George Hack put in hand the design of a more substantial big end and by the time of the Ulster Grand Prix, Graham Walker had his foot-controlled lubricator fitted.

A win by Graham in the Dutch TT at 75.96 mph (122.32 km/h) brightened the scene a little for the Rudge team members. Then, on June 30, H.G. Tyrell Smith, who had finished fourth in the Senior TT, scored a fine win in the 500cc Class of the 25-mile Irish Road Racing Championships in Phoenix Park, covering the distance at an average speed of 65.5 mph (105.5 km/h).

At the Nürburgring on July 8 the Rudge team rode in the 15-lap (424 km) 500cc Class of the German Grand Prix and Graham came second to Charlie Dodson on the Sunbeam, with Ernie Nott third and Ashby sixth. Despite the fact that not all of the desired TT improvements had been made to their machines, they were definitely more reliable now.

At the Grand Prix d'Europe held over the Meyrin Circuit near Geneva at the end of July, Ernie Nott and Graham Walker finished second and third respectively in the 250-mile (402.6-km) 500cc Race. Since this was behind the brilliant Wal Handley and his Dougal-Marchant-tuned Motosacoche, they were not doing too badly. Ernie also finished second in the 600cc sidecar event at the same race meeting.

Then, as a prelude to the Ulster Grand Prix, Graham ran in the well-known Irish handicap event, the Leinster '100', and scored a brilliant win despite strong opposition from a certain Stanley Woods on his 490cc Norton.

On Saturday, September 1, 24 riders and their machines lined up for the start of the 500cc Class of the Ulster Grand Prix over the 20½-mile (33-km) Clady Circuit. Amongst them were the four Rudge-Whitworth teamsters, Graham Walker, J.G. Tyrell Smith, Ernie Nott and Jack Amott. Complete with his foot-oiler now, Graham was ready to take on Charlie Dodson and his Sunbeam, who had two great natural advantages over him anyway. Firstly, Dodson was of much slighter build and lighter weight and, secondly, Graham still suffered from a wound he had received during World War 1 that restricted his ability to use his left foot. It would be interesting to see whether the performance and newly acquired reliability of his machine would compensate for these.

On the first lap Dodson (Sunbeam), the previous year's winner, lapped at 78.59 mph (126.55 km/h) from a standing start, followed by Tommy Bullus (Raleigh) at 78.01 mph (125.62 km/h) and Graham Walker on his Rudge at 77.84 mph (125.35 km/h). The Dodson-Walker duel had now started. Next time round they were riding wheel-to-wheel, the Irish crowds cheering them on, with Graham possibly a fraction ahead.

Charlie refuelled on Lap 4, taking only 30 seconds to do so, and tied for the lead with Graham at 80.04 mph (128.89 km/h), Bullus on the Raleigh was a minute behind in third spot while Tyrell Smith was fourth. Graham refuelled on the next lap, then Dodson did likewise two laps later and Graham on the lap after that. Amazingly both their motors were surviving the killing pace! They 'dead-heated' again on total running time on Lap 8. Who was it going to be? Who was going to win this monumental struggle? The crowd was wild with excitement!

At the start of the last lap, Charlie Dodson held a

Bob Dicker (499cc four-valve Rudge-Whitworth) after winning a three-lap handicap at over 91 mph during the course of the 1928 Brooklands' 'Hutchinson Hundred' Race Meeting.

narrow 70-yard (65-metre) lead, but Graham Walker seemed to have the greater speed on the straights and was coming up fast again. Would he beat Charlie? Every eye followed the two riders' indicators on the scoreboard at the pits. Dodson led by a mere 10 yards (9 metres) at Muckamore! At Clady they were signalled as being together. Fantastic racing! Only a mile to go now!

The Rudge-Whitworth pit team members were now sweating blood in anticipation! Had Graham done it? Yes, yes, he had! He topped the horizon and soon screamed over the finishing line a clear and decisive winner some 11.4 seconds before Charlie. His incredible last lap of 85 mph (136.89 km/h) broke the previous lap record by almost 5 mph (8 km/h), a truly magnificent performance.

Graham Walker won at a record average speed of 80.78 mph (130.08 km/h) with Tyrell Smith third at 78.08 mph (125.73 km/h), Ernie Nott fourth and Jack Amott seventh. In so doing he not only confirmed the Ulster Grand Prix as being then the fastest road race in the world, but also became the first rider to win an international road race at over 80 mph (130 km/h).

Meanwhile, earlier, at home, Ernie Nott had managed to persuade George Hack to enter one of the new four-valve racers at Brooklands Track in the 500cc Solo Race staged as part of the BMCRC's annual 200-Mile Solo Races for all classes, held that year on Saturday, June 23. The avowed objective he put to George was that of proving the value of the new redesigned big end, but he really had a bigger target in mind. He wanted to gain sufficient track-riding experience to enable him to have a try at the world's 500cc solo two-hour record. During practice for the event his lack of track-craft and frenetic riding caused much head-shaking and comment amongst old track hands. In the event itself, however, the sceptics were well and truly confounded for he came home a fine third at 89.24 mph (143.75 km/h), Bert Le Vack (496 New Hudson) winning from Bill Lacey (498 Grindlay-Peerless-JAP) by a commanding three laps.

His next appearance at the track was at the BMCRC's 'Cup' Race Meeting on Saturday, October 6. He had clearly learnt from his previous riding experience at Brooklands, for although he did not manage to win any of the events he did succeed in gaining one of the coveted Brooklands' Gold Stars with a lap at 100.27 mph (161.47 km/h). Ernie's machine was now performing so well that George Hack agreed to him going for the 500cc class two-hour world record. The attempt was made on Friday, October 19, practically without any preliminaries, and Ernie proceeded to reel off an amazing series of laps at, for the day, an amazing speed, and kept on doing so, thereby deflating his critics.

Starting off with a lap at 87.84 mph (141.45 km/h), he worked up to an average of 102.90 mph (165.70 km/h) for his 15th circuit and thereafter his slowest laps were covered at 99.41 and 99.81 mph (160.08 and 160.72 km/h) respectively due to the presence of a car on the track at the time. After 37 laps had been completed or almost half-distance, he stopped for a quick fill-up, but so rapid were his pit workers that only 50 seconds were lost in topping up with fuel and oil.

He took the two-hour record for Classes C, D and E (for 500, 750 and 1000cc solos) easily, covering 200 miles 816 yards (322.81 km) in that time — an average speed of 100.23 mph (161.40 km/h). The 200-mile record also fell to him in a time of 1 hour 59 min 43.15 sec. The previous two-hour records had stood to Bert Le Vack (in Classes C and D) and Claude Temple (in Class E) at 98.40 and 95.66 mph (152.65 and 154.04 km/h) respectively. Temple's record had been set up on the French speed-bowl at Montlhéry, which was a good deal smoother than Brooklands. It is worth noting that Nott's final laps were covered in a heavy shower of rain yet, despite this, his pace continued unabated. This was a truly magnificent achievement and a fitting climax to a fine year of competition success for the Rudge-Whitworth concern.

But there were many other successes that had come to the marque during 1928. Earlier in the year A. Wilcock (Rudge-Whitworth) had won the premier award in North-Western Centre ACU's Championship Trial and had earned the title of 'Solo Champion of the North-West Centre for 1928'. Meanwhile, in Holland, a Rudge rider was amongst the only six to finish out of 70 riders who started in the notorious 1100-km trial staged on May 4 and 5. Then, on May 20, a Rudge rider set up fastest time of the day and broke the track record in the Hilversum Grass Track Races after having also set a new track record on the Amsterdam Grass Track.

Further afield, M.R. Matthews scored numerous racing successes in Egypt on his Rudge-Whitworth including 14 firsts, four seconds and the breaking of two course records. Other Continental successes included the first prize in the Friedland Annual Reliability Trial, and first and third in the Amateur Class of the Sappemeir Grass Track Races. In June 1928, a Rudge-Whitworth Special was taken to victory in the Polish Tourist Trophy Races, making fastest time of the day in the process.

The year 1928 had seen the emergence of another popular branch of motorcycle sport, dirt track or speedway racing as it is known today. At first only tuned standard or sports machines were raced, but it soon became obvious that there was a market here for special models and, in June 1928, the first Rudge-Whitworth speedway machine was assembled. The parts used were mainly from existing production models. A 1927 works-type competition engine was fitted into a standard frame. The wheels had 1927-type brake-less hubs and 1928-type rims. It was decided to stiffen up the frame by adding two struts running from the steering head to the rear wheel spindle. These modifications proved a bad mistake, though, and in one test lethal. During the test in question, the rigidity of the strutted frame prevented the rider from laying the machine over for a bend: as a result he was thrown off, hit a wall and died from his injuries.

Despite this tragedy, the machine performed well and went into limited production in June 1928. Early in July the first of these cinder racing machines was put on display in the Tottenham Court Road showrooms of Glanfield Lawrence Ltd, the London Rudge-Whitworth agents. It had a fixed gear of 8.0:1 and was priced at £70.

The first success for the new model came at the opening of the Lea Bridge Track the same month, in the hands of 'Buster' Frogley. In the Essex Scratch Race he took his nickel and scarlet Rudge to victory against T. Croombs (Peashooter Harley). It is interesting to note here, that the seventh of these speedway Rudge-Whitworths to be produced that year was bought by a woman rider, Fay Taylour, who was making quite a name for herself on the cinders.

Attention was now switched to the models for the forthcoming (1929) season. All three models in the 1928 Rudge-Whitworth range were of around 500cc capacity. Some variety was needed in the range to attract prospective new customers. The '350' was a popular size of machine so, in July 1928, a new 339cc Rudge engine was designed as an interim measure. Apart from the adoption of redesigned valve gear and the use of radially disposed exhaust valves (as with the 1928 Special and Sports Special machines) plus dimensions of 70 mm bore and 88 mm stroke, it was merely a scaled-down '500' power unit. The rest of the machine made use of the larger-capacity machine's cycle parts and gearbox.

The rapidly expanding demand for lightweight machines of 250cc and under was not lost on John Pugh and it was decided to develop quickly models of that capacity. With the amount of time required by the design and development department to devote to improving the 499cc racing models for the coming season, there was none left to design a completely new lightweight power unit and so a proprietary one was sought as a 'stop-gap' measure. First, experiments were carried out with a 250cc Villiers two-stroke engine, but this had insufficient power to provide the required performance and so was abandoned. Instead it was decided to experiment with lightweight JAP engines and, on October 15 1928, two prototype machines were built, one powered by a 249cc (64.5 × 76 mm) side-valve unit and the other by a 245cc (62.5 × 80 mm) overhead valve engine of that marque. Tests proved satisfactory and these two machines were put on display at

Roger ('Buster') Frogley astride his Dirt Track Rudge at the Wimbledon Track in 1928.

the Olympia Motorcycle Show in November. Meanwhile the company had rechristened its Sports machine the 'Ulster' to commemorate Graham Walker's splendid victory. The Rudge-Ulster was to remain in production for the next ten years and gain a high reputation as a quality road-going machine for the sporting rider. In its production form, the Ulster had several refinements, including totally enclosed overhead valve gear lubricated by oil mist from the crankcase and a primary chain enclosed in a cast, aluminium alloy case. Its price, at £69, was only £1 less than the speedway machine, which was christened the 'Dirt Track Special'. The 350's price was set at £49, while the road-going 70-mph (112-km/h) chrome-plated '500' special was £6 dearer. The new lightweights cost £45 for the ohv and £39 10s for the side-valve. Their specifications included Rudge-Whitworth four-speed gearboxes, 6¾-in (171-mm) diameter coupled brakes and 2½-gallon (11.4-litre) petrol tanks. A simplified design of lightweight Rudge fork was used on these models, while their saddles were made by Lycett as on the larger machines.

A feature of all of the 1929 models shown at the Olympia Exhibition was the standardisation of electrically equipped models at a slight premium, ML Maglita combined ignition and lighting sets being employed. Detailed improvements included neater carriers, more efficient mudguarding and more silent valve gear, while a neater form of speedometer mounting was employed, in which the instrument was carried on the front fork. A number of new sidecar designs were also listed. Experiments had been carried out with an enlarged, 595cc, engine for sidecar work in June 1928 but the project was not continued with. Otherwise such a model might well have been featured for the first time in the firm's 1929 motorcycle catalogue.

10

Things start to go right (1929)

A formidable record of sporting success had been gained by the Rudge-Whitworth marque by 1929 and its riders had become a considerable force to be reckoned with both in reliability trials and road racing. The year started well for the firm with numerous successes in British trials events, culminating in the summer of 1929 with top performances in both the Scottish and International Six Days' Trials.

Eight Rudge-Whitworth riders took part in the 'Scottish' in May that year, namely: C.R. Sanderson, a well-known speedway rider of the day; Fred Povey; Jack Amott, the road racer; J.T. Sleightholm; Jack Williams; H.S. Cussens; Geoff Butcher and J.B. Ireland. Of these, Jack Amott made the best solo performance of the whole trial and Geoff Butcher the best sidecar run. In all, Rudge riders gained five silver cups and two gold medals.

The result of all these successes was that the Rudge trials' man, Geoff Butcher, was invited to join the British Team to take part in the International Six Days' Trial to compete for the International Trophy. In this event, which started on August 26 at Munich in Germany and ended six days later at Geneva in Switzerland after passing through five different countries, he completed the course with the loss of no marks. He was awarded a gold medal, and the British Team of which he was a member, and which lost only a single point, was presented with the International Trophy. The Rudge-Whitworth No 2 team, comprising Jack Williams, Fred Povey and G. Pycroft, won the Manufacturers' Team Prize in the 500cc solo class, while Geoff Butcher, C. Edge and R. Neisse, gained the Manufacturers' Team Prize for the firm in the 600cc sidecar category, in which they were the only manufacturers' team to finish without any of its members losing any marks.

All was by no means plain sailing in this event, though, for on the last night of the trial a protest was lodged against Geoff Butcher because his sidecar passenger was a German national. The FICM regulations, however, permitted the carrying of ballast and the passenger was regarded as such, so the protest failed! Also, on the last day of the trial Geoff Butcher nearly put the British team's chances in jeopardy when his throttle wire broke. Fortunately he was able to carry out a repair so quickly that he lost no marks for lateness at his next check point.

Five riders and their machines made up the Rudge-Whitworth works team for the 1929 Senior TT Race in the Isle of Man: Graham Walker, H.G. Tyrell Smith, Ernie Nott, Jack Amott and T.W. Denney. A number of design changes had occurred with the machines and in combination these were to provide both increased reliability and speed.

Up to and including 1928, various types of wet sump lubrication, involving metered, constant-loss pumps, had been employed. For the 1929 races, however, a new worm-driven double-acting pump working in a pair of bushes in the timing case had been designed, to provide dry-sump lubrication. It drew oil from an oil tank mounted on the saddle pillar and delivered it via the mainshaft through a hole drilled in the offside flywheel to the big end. To avoid weakening the crankpin, the usual practice of drilling it was avoided. Instead, the oil found its way to the bearing surfaces through a groove cut on the circumference of the pin. A bypass took about 16 per cent of the oil supply and delivered it to the rear cylinder wall. A scraper at the back of the crankcase removed the oil from the flywheel and dropped it into a sump situated below and behind the crankcase, from where it was drawn up through a channel in the wall of the crankcase and through the timing case to the top end of the pump. From there it was pumped back to the oil tank again.

The crankcase had also been redesigned. Previously it had had a square flange with four studs, but this was inadequate for the 7.25:1 compression ratio envisaged and finally used. The base was therefore made round, which allowed the use of six retaining studs for greater security. A further crankcase modification involved the addition of a second drive-side bearing, the housing of which was reinforced with eight strengthening ribs — two of these being eventually dropped when this modified crankcase was used on the 1930 production models.

So far as the cycle parts were concerned, a longer, better-handling frame was now used and a rocking-pedal type of foot-operated positive-stop gearchange fitted on the nearside of the machine — except on Tyrell Smith's model which retained the previous handchange arrangement. Furthermore, the nearside handlebar was now fitted with a lever to control the steering-damper adjustment. This avoided the hazards associated with removing a hand from the bars at high speed, possible under conditions of incipient high-speed wobble.

The new footchange had been largely developed by Tyrell Smith, who had tried it out first in the North-West '200' road race, a few weeks earlier, when it gave him a lot of trouble. This experiment so prejudiced him against it, that he was the only Rudge team member to opt for staying with the handchange system.

The first week of practising for the TT Races started off well for the Rudge team, with Ernie Nott gaining the third fastest 500cc lap time on the first day and Graham Walker heading the practice leader board on all the other days except the Friday. Despite this, though, the team had a major problem to contend with; the cylinder heads of their engines were soft! Each day of practice, the machines would start out with plenty of compression, only to come back with none, the valves sunk hopelessly in their seatings and pushrods bowing in the middle beneath the pressure of of springs and tappets. Graham Walker completed a record 28 practice laps before the problem was solved. This occurred with the arrival of new cylinder heads a mere 48 hours before the time to hand the machines in for weighing and official examination before the race. Anyway, it put the Rudge team members out of their misery. According to Graham: 'Our garage looked for the world like a hat shop after a pernickety customer had spent an hour or two looking for a hat the shopkeeper hadn't got. There were scrapped cylinder heads piled in tiers and, believe me, by race day we could have taken on the world at "top overhauls"!'

The race itself was notable for the fact that the Rudge team was now really in the picture for the first time since its TT win in 1914. When the morning of the race arrived, the hearts of the competitors sank — it was raining. The only consolation was that with the exception of the stretch from Quarter Bridge to Ballacraine, the circuit was now covered in non-skid tarred chippings. This exception was to result in one fatality and a number of hair-raising experiences, not least for the Rudge riders.

The greasy section of the course first started to play its deadly role in the proceedings in the first lap. The worst spot was where machines were heeled over for the third-gear sweep through the treacherous Gleeba Bridge. First to come to grief was Wal Handley on his AJS, who blared into the corner only to crash and slither helplessly into the wall. He had barely time to drag his battered machine out of the way before Lamb (Norton) met his death at the same spot. Then Jack Amott crashed his Rudge, breaking his collarbone and arm. Then, as Wal rushed over to try and rescue him, four other riders, including Ernie Nott and Graham Walker on their Rudges, arrived in quick succession, skating and slithering around the struggling men.

Amazingly, despite all this, the favourite to win, Tyrell Smith (Rudge) — who managed to avoid getting involved in the Gleeba Bridge mêlée — led at the end of this horrendous first lap having raised the lap record from a standing start to 71.88 mph (115.75 km/h), almost 2 mph (3 km/h) faster than the previous year's fastest flying lap. His team mates, Graham Walker and Ernie Nott, lay second and third with standing start laps of 71.16 and 70.57 mph (114.59 and 113.64 km/h).

Although Tyrell and Graham came in to refill at the end of the second lap, the former still led, having averaged 71.90 mph (115.78 km/h), while Graham had cut back his lead to 18 seconds or so. But Dodson (Sunbeam) and Hunt (Norton) had both put on a spurt and relegated Ernie Nott to fifth spot. Charlie Dodson had broken the lap record in the process at 72.93 mph (117.4 km/h), but Tim Hunt had gone even faster to lap at 73.12 mph (117.75 km/h) and take the lead ahead of him. The fastest Rudge, with Tyrell Smith in the saddle, now lay third, while Graham Walker was right back in sixth place — but only 43 seconds behind the race leader! The crowd was now going mad with excitement! What a race!

Then the Isle of Man 'gremlins' decided to have a go at the race leader, Tim Hunt, who hit the wall on the outside of the right-hand turn at Quarter Bridge and fell, losing a footrest and some skin in the process; quickly recovering, he got on his way again though by now he had dropped back six places. Then Tyrell Smith came a purler on the left-hand sweep of Craig Willies, which abraded away a large proportion of his green corduroy riding breeches! This dropped him back to seventh spot, 12 seconds behind Tim Hunt on the Norton. Dodson was now comfortably in the lead on his Sunbeam.

Graham Walker now had things under control, with his Rudge going well, but it was lap 5 and the Manx gremlins were all set now to get their teeth into the Rudge-Whitworth team. Rounding the right upper sweep of Craig Willies, Graham suddenly started to broadside his machine on the wet road and then, to his horror, he saw a semi-conscious Ernie Nott squatting, gazing down the hill, with his recumbent Rudge lying right across the road. Suddenly 'Nottie' regained full consciousness and made a flying leap for the safety of the bank at the side of the road whilst trying to drag his machine with him. Graham collided with the back wheel of Ernie's machine, but managed somehow to stay on board, with the knowledge that he still had 30 seconds to haul back to get into the running again. Then 12 miles (19.3 km) further on, while coming round the White Cottage bend just before the Ramsey Straight, nemesis struck. Graham's engine suddenly went dead and he received a blow to his cheek. A quick look down revealed a missing exhaust tappet and a pushrod which had broken in half. Subsequent analysis of the latter, after the race at the National Physical Laboratory, revealed that it had a faulty grain structure. What rotten luck! But there was more to come for the Rudge team, for on that same fatal lap, Tyrell Smith had his second crash, this time just outside the famous Glen Helen Hotel. It resulted in a broken rib.

Years later, when asked to recount what had actually happened, he said: 'I took it a shade too fast. The exhaust pipe touched the road and spun me round. I crashed into the bank, was winded and my leather jacket badly torn. They carried me into the hotel and pinned my clothes together.

'After a bit I got my breath back but had a nasty pain in my chest. However, I came out of the hotel to see what was doing and found that someone — Bob MacGregor, I think it was, but I was never quite sure for I was still rather dazed — was holding my machine up with the engine running and the clutch out, strictly contrary to race regulations concerning assistance to riders en route, all ready to move off! They sat me on it and, still very muzzy, I let in the clutch and carried on.'

Carry on he certainly did, for despite his injuries, which forced him to stand on his footrests to ease the pain and a further crash at the Gooseneck, he carried on grimly and by sheer guts moved up into fourth place. Tim Hunt meanwhile had recovered from his fall to move into second spot, while Ernie Nott, who had restarted after his earlier mishap at Craig Willies, now lay fifth on his slightly 'bent' Rudge.

Hunt (Norton) was overtaken by Alec Bennett (Sunbeam), Charlie Dodson's team-mate, and the time difference between him and Tyrell Smith steadily got less and less. On the last lap the plucky Irishman really opened up the taps despite his pain, overtaking Hunt in time to take third place for Rudge-Whitworth at 70.25 mph (113.1 km/h). Ernie Nott maintained his fifth position to the finish, averaging 69.80 mph (112.40 km/h) for the race.

The event certainly proved one thing to the Rudge team. It was no good having fast machines, good reliability and brilliant riders, if the 'Manx luck' was against you.

Ernie Nott seemed to have an affinity with Brooklands and had learnt his track craft well now. He decided to have another crack at the 500cc Class of the Brooklands 200-Mile Race on Saturday, July 27 1929. John Pugh and George Hack gave the go ahead. A win in that event would certainly do much to restore the sagging Rudge racing fortunes and the sagging morale of the Rudge works racing team.

On the day, Ernie was highly successful, dominating the race from start to finish. The only man to come anywhere near catching him was Bill Lacey and his 498cc Grindlay-Peerless-JAP; even *he* was lapped by him though. Ernie won the event at the fine speed of 100.07 mph (161.14 km/h) — the first and only occasion on which the Brooklands 200-Mile Race had been won at over the 'ton'. Much of the credit for this machine's performance can be ascribed to Jim Dalton, the Rudge test-bench engineer who, together with Ernie Nott and under George Hack's direction, spent many hours extracting the last bit of power out of its engine.

The 'iron man' of motorcycle racing in the late 1920s and early 1930s, Ernie Nott, at the start of the 1929 Brooklands' 200-Mile Race, which he won at 100.07 mph — the first and only occasion on which that event was won at over the 'ton'.

After the various TT disasters, things started to brighten up for Rudge-Whitworth on the road racing front. On Sunday, June 30, despite still suffering from the injuries he received in the Isle of Man, Tyrell Smith finished first in the Czechoslovakian Grand Prix. A week later, on July 6, the Rudge team was nearer to home, in Holland, for the Dutch TT and this time it was Ernie Nott who was in winning mood, with Graham Walker second. Ernie's winning speed of 75.75 mph (122.00 km/h) was a new race record.

The last Sunday in July saw the team at the Nürburgring for the German Grand Prix. Once again Tyrell Smith set a cracking pace in the 500cc Class, breaking the course record and winning at a record average speed of 63.38 mph (102.1 km/h), an amazing 6 minutes ahead of the second man, the Australian rider Arthur ('Digger') Simcock and his Sunbeam. Graham Walker, the next Rudge rider to finish, came home into seventh place.

Of course, being the track-record holder and previous year's winner, Graham was looking forward to the Ulster Grand Prix — his favourite race of the year — with guarded optimism. The result of the Isle of Man Senior TT, it was felt, might have been different if Graham had not had a pushrod break in his fifth lap. Anyway, between the TT in June and the Ulster Grand Prix in September, several more horsepower had been extracted from his engine. His optimism was thus well founded.

During the practising period for the Ulster it was essential to get the third gear right for the uphill portions of the undulating 7-mile (11.2-km) Clady Straight. One morning while out testing a slightly higher gear than usual, Graham Walker whistled over the brow of a hill on the renowned straight to find himself face-to-face with an enormous cow! It was the first of a herd of 37 about to undergo early morning milking. He nearly had a heart attack! After that little episode the morning experiments were terminated and an appointment made with a large flask of brandy, post haste!

Race day, Saturday, September 7, turned out to be beautifully sunny and by 11:30 am, when the 68 starters left the paddock, had turned into a real scorcher. The Rudge teamsters were mentally calculating where on the course the patches of molten tar were likely to be and how to avoid them! Their plan for the race was a simple one. George Hack, the team chief, had issued instructions that so long as none of the opposition overtook Graham Walker, he was to lead the Rudge trio. In consequence, after a good start, Graham was a bit surprised — to say the least — to have Ernie Nott overtake him at Ballyhill Corner on the first lap.

Graham gave him the 'slow down' signal as he wanted to warm his motor up gently before giving it the gun, but Ernie rejected this advice with a shake of his head. Twice more before reaching Thorn Cottage he caught up with Ernie and got the same response. Then, on the Clady Straight, all became clear, when C.J. Williams crept past both of them on his Dan O'Donovan-tuned Raleigh, his engine revving like a band-saw. Ernie, fighting his way through the massed start, had realised that some of the opposition were faster than anticipated! Past the Grandstand Williams led the Rudge pair by some 50 yards (46 m) to complete his standing-start lap at an amazing 81.18 mph (130.7 km/h).

On Lap 2, approaching Ballyhill, Williams, Walker and Nott were all in a bunch fighting for the lead when the Raleigh broke its magneto driving chain and suddenly went dead. But this did not ease the pressure on the two Rudge men since Stanley Woods was now just astern of them on the works' Norton. No time to relax now. As a result, the second lap was even faster than the first, with Graham Walker in the lead past the Grandstand to complete it in a new record time equivalent to 82.36 mph (132.62 km/h) some 30 yards in front of Ernie Nott. The race average of the leader had now gone up to 81.68 mph (131.53 km/h).

It was on this lap that Graham had a most unpleasant experience. While riding flat out down the notoriously bumpy Clady Straight, he realised his eyes were going out of focus — instead of a white road between green banks, all he could see was a sickly yellow streak in a sea of greyness, and all he knew was that he must stop on that streak somehow. Shaking his head and looking at the roadside cured the trouble, but it promptly returned immediately he got down to it again. It was then that the explanation dawned on him. In the 1928 race he had no rear mudguard pad and had ridden most of the race with broken saddle springs. This year (1929), determined to avoid the same appalling discomfort, he had organised a somewhat luxurious pneumatic rear cushion so that he could 'lie down to it' in comfort. He was flatter as a result and had to cock his head further back in order to see the road ahead. This had the effect of temporarily restricting blood flow to the brain, causing a temporary blackout. Later he found that other riders had experienced the same problem as well. Lap 2 also saw Tyrell Smith stepping off his machine to avoid another, fallen, rider.

Despite all these problems the end of the third round saw Ernie Nott and Graham Walker running past the Grandstand, wheel-to-wheel, with Stanley Woods (Norton) right on their tails. At the end of the fifth circuit the leading pair both came in to refill their tanks and get away again smartly before Stanley could catch them. Lap 6 saw Graham with a 200-yard (185-m) lead with Ernie and Stanley battling, wheel-to-wheel, for second place. Meanwhile, Tyrell Smith, having remounted his machine, had steadily caught up and was now running fifth.

On the seventh circuit, Stanley Woods managed to overtake both Ernie Nott and Graham Walker, just before Clady Corner. Then, as the three riders rushed down to the bend, Graham was on the outside of Stanley, whose foot slipped off his brake pedal, and was obliged to follow him straight towards a massive grass bank — which they both managed to avoid hitting through deft footwork. Ernie, seeing his chance to take the lead, tried to nip round behind them, but overdid matters in his excitement and dropped his machine with a resounding crash. By the time he had picked himself up and got going again, Stanley and Graham were passing the Grandstand again, almost shoulder-to-shoulder.

During Lap 8, the two men constantly swapped the lead, first Graham Walker and then Stanley Woods being in front. At the end of that circuit the timekeepers again recorded a dead heat! Ernie Nott, as a result of his fall on Lap 7, was now running 21 seconds down on the two leaders. Nearly twisting the throttle off the handlebar in his efforts to catch up in the ninth circuit, he shot through the Colonial Section like greased lightning and then in the act of passing A. Greenwood (New Henley) on the crest of a bump, disaster occurred. Both machines took off, the riders colliding in mid-air. Greenwood crashed, but was unhurt. Ernie Nott, however, was thrown over a wall and received a huge horseshoe-shaped cut on his forehead, his helmet having been torn off by the impact. He was taken to hospital, but Ernie was not known as the 'iron man' of racing for nothing, though, and was soon able to be discharged little the worse for wear.

Meanwhile, Stanley Woods (Norton) was having trouble sorting out his gears, on Lap 9, and fell while getting the wrong gear. Remounting, he then had to come in for a pit stop to fill up his tank. The result of all this was to give Graham Walker, who now led, a chance to ease up a bit. Despite this Graham averaged a fine 81.04 mph (130.50 km/h) by the end of the tenth circuit. He was now plagued by the terrible heat, melting tar on the roads and hordes of flies determined to commit hari kiri on his goggles.

Graham was now so far ahead that the overtaking of the unfortunate Stanley Woods by the latter's team-mate, Tim Hunt, was unlikely to provide a threat to him. Graham's last lap was therefore fairly uneventful and he came home a relatively easy victor at 80.63 mph (129.8 km/h) to a terrific ovation from the Irish crowd and a very welcome glass of Bass handed to him as he pulled up some 200 yards (185 m) after the finishing line. Tyrell Smith, the next Rudge rider to finish, brought his somewhat battered machine home in fourth place.

Later in the year, on Sunday, October 20, Graham Walker took part in the FICM Grand Prix in Spain, held over the Ametila Circuit. On that occasion the finishing order was reversed, with Tim Hunt taking the honours on his Norton.

Private owners overseas were also having competition successes riding Rudge-Whitworth machines and on Sunday, June 23, Terzo Bandini (Rudge-Whitworth) gained a second place in the six-lap 136-mile (219-km) Italian Grand Prix staged over the Lario Circuit. The same day, at the grass track races held at Alkmaar in Holland, Rudge-Whitworth riders won most of the important events: A.P. van Hammersveld winning the 500cc Senior Race and A.W. van Dinter the 500cc Junior event; van Hammersveld also set up a new track record. At the Kolberg Race Meeting in Germany, Erich Tennigkeit drove a Rudge-Whitworth sidecar combination into second place in the 600cc Sidecar Race.

Moving further afield, in South Africa, while Malcolm Campbell was attempting to set up new world speed records with his *Bluebird* car on the Verneuk salt pan, C. Collins, riding a 499cc Rudge, established a new local South African flying-start kilometre motorcycle record with a mean speed of 91.23 mph (146.9 km/h). On the same machine, J. du Toit covered five miles at 91 mph (146.5 km/h) despite a strong cross-wind. Prior to this, on July 28, some record-breaking activity had taken place at Melbourne in Victoria, Australia. Jimmy Pringle, riding a 499cc Rudge-Whitworth tuned by Alan Bruce, later of HRD fame, had set up a new mean speed Australian National Record for the flying-start quarter mile (402.5 m) in the 500cc Solo Class at 98.91 mph (159.3 km/h), with one-way speeds of 100.00 and 97.82 mph (161.0 and 157.5 km/h) respectively in the two opposed directions.

In August, Basil Nixon, a Rudge-Whitworth rider living in Melbourne, started out on his successful attempt at setting up a new record for the circumnavigation of Australia. Starting from his home city, his route encompassed Perth via Adelaide in solo trim, then with sidecar attached to enable extra supplies of water, petrol and oil to be carried, he was able to tackle the 1600-mile (2576-km) run to

Broome, via Meekatharra and Nullagine. Thence from Broome to Darwin in the Northern Territory and from Darwin to Brisbane, a distance of some 2500 miles (4025 km). The final stage back to Melbourne from Brisbane included Sydney and Canberra, giving a total distance for the whole trip of 9000 miles (14,492 km). Then, in September, closer to home, Faura and Segura gained first and second places respectively in the Spanish Championship Races at Penya Rhin near Barcelona.

Also in September, Rudge-Whitworth Ltd announced three of its models for the 1930 season — two 500s, the Ulster and the Special, and the latest 350. All three had new design features. Their frames were generally much the same as on the 1929 models, though. All three engines were, of course, of the overhead four-valve type and on each model a four-speed gearbox and coupled brakes continued to be used.

Perhaps the most outstanding change in design concerned the rocker gear, which on the 500s was now supported upon standards cast integrally with the cylinder head. The four uprights carried lighter rockers than hitherto, and their bearings were now mounted inside the standards themselves and not inside the rockers as on the previous arrangement. The side-by-side pushrod arrangement, with the rods working on inverted cups on the rockers, was retained, while a neater rocker-box cover and tube to enclose the pushrods greatly enhanced the appearance of the engine. The valve gear of the 350 remained unaltered from the 1929 design, the rockers being supported on plates bolted to the head. The four valves on all machines each had two springs.

Dry-sump lubrication was now featured on all these models. The system employed was similar to that used on the works racing machines in the 1929 TT Races, but had been considerably 'cleaned up' in its design. The various pipes with their banjo unions and the 'tell-tale' indicator were now much more neatly laid out.

All the engines now had heavily ribbed crankcases, two-ringed, dome-top, aluminium-alloy pistons and H-section connecting rods with roller bearing big ends. The crankshafts were carried on three bearings, ball and roller types being used on the driving side and a roller bearing on the timing side. Six bolts held down the cylinder barrel to the crankcase, the barrel flange being circular.

The Ulster engine had its sparking plug central in the cylinder head, a section of the petrol tank being cut away on the nearside to enable easy access for the plug spanner. The Special on the other hand, had its sparking plug located between the bases of the nearside rocker pillars. This model was in fact a touring edition of the Ulster, the plug position being the only obvious difference in its engine design. The Ulster, of course, had a specially tuned engine, its normal compression ratio being 7:1 as against 6:1 for the Special. A further distinguishing feature was that the Ulster had a separate oil tank mounted on the seat pillar, whereas the 350 and the Special carried their lubricant in a compartment in the petrol tank. The latter on the Ulster model thus had an increased capacity, compared with the Special, of 3 gallons (13.6 litres), while its separate oil tank was able to hold 3 pints (1.7 litres). The chains on the Ulster were again enclosed in a cast aluminium-alloy case and guard.

Rudge gearboxes had also undergone improvement for 1930 and were claimed to be 50 per cent stronger than the 1929 versions in consequence. The teeth of the gearwheels were now of wider pitch, continuous and not grooved, as hitherto, for the selector forks. Instead the grooves, being cut on the sides of the wheels, did not now divide the teeth. The kick-starter lever had also been redesigned, the pedal being hinged to fold parallel with the machine when out of use, out of the way of the rider's leg. The solo gear ratios selected were as follows: the Ulster, 4.4, 5.4, 6.9 and 11.9:1; the Special, 4.6, 6.3, 8.3 and 15.2:1; and the '350', 5.5, 7.4, 9.7 and 17.5:1. The corresponding values for sidecar work were: the Ulster, 5.2, 6.4, 8.2 and 13:1; the Special, 5.2, 7.2, 9.4 and 17.3:1; and the '350', 6.0, 8.3, 10.9 and 20:1.

The wheels of the Ulster and the 350 were shod with 27×2.75 in (686×70 mm) tyres, while the Special carried 26×3.25 in (660×83 mm) covers. Alternative sizes could be provided for all three machines at slightly extra cost. The wheels were, of course, interchangeable on all models, and the braking system — the well-established Rudge proportionately coupled method of operating both brakes by pedal with an independent control for the front wheel — remained unchanged. The 500s had brake drums measuring $8 \times 1\frac{1}{2}$ in (203×38 mm), while those on the 350 were $6\frac{3}{4} \times 1$ in (171×25 mm).

Amongst the features common to all three models were the ML 6-V Maglita lighting sets, with an improved generator, and the latest design of Miller headlamp incorporating an ebonite switch and a larger ammeter dial, both on top of the lamp housing within easy reach of the rider. The battery was carried in a sturdy and novel clip bolted direct to the seat tube on the Special and the '350', but to the chainstays on the Ulster. The carrier incorporated a baseplate which made it almost impossible for the battery to slide out of the clamp.

On all models the tank-side-located, gear-change lever was fitted with a large wooden knob; while the handlebars, which were adjustable, had been modified so that the bars did not foul any sidecar that might be fitted.

The shock absorbers and a steering damper were, of course, embodied in the forks while the shackles were adjustable. A marked departure was the adoption of almost semi-circular cross-section mudguards, that fitted closely to the tyres. Not only did these prove more effective than the former type, but they also greatly enhanced the machines' appearance. The detachable flap on the rear mudguard, provided to assist wheel removal, had been strengthened with mudguard stays to prevent vibration.

An interesting fitting was the detachable rear carrier, which took the form of a pressed-steel grid, which could be instantly fitted (or removed) by means of four clips to a pair of handrails. On the Ulster the toolbox was clipped to these two rails, but on the other two models it balanced the battery on the seat pillar. Also, the spring-up central stand had been strengthened.

The remainder of the equipment included adjustable knee grips, a Lycett Aero saddle, a Tecalemit grease gun and a tool kit. The standard finish was black with gold lining and lettering, but an optional finish of black with maroon mudguards and tank panels could be obtained at no extra cost. Chromium plating could be supplied at extra cost on all parts hitherto nickel plated, including the 'Brooklands'-type silencers. The tank panel could also be chromium-plated to order at extra cost.

How much did the standard models cost? Well, including the electric lighting system, the prices of these 1930 models were set at: £52 for the 350; £59 for the Special; and £70 for the Ulster. For another £5, the Ulster could be obtained guaranteed to do 90 mph (145 km/h) and for a total cost of £85 it could be supplied tuned to do 100 mph (161 km/h); certificates being supplied to these effects. This certification of having exceeded certain speeds on these latter versions of the Ulster involved the works racing team in many hours of extra riding. Sometimes machines certified to achieve a given speed would be returned by dissatisfied owners to the factory, saying that their particular model could not achieve the guaranteed speed. As Graham Walker was to recount in later years: 'this was due, as often as not, to a lack of effort on the part of the owner as much as anything else.' The owner would be invited to demonstrate his machine's alleged lack of performance over a selected course laid down over public roads, on which he would be timed.

Almost invariably a works rider would achieve the requisite performance over the same circuit on the same machine, finishing with his engine well-nigh red-hot and the bottom of his footrest rubbers almost bare due to contact with the road on the bends! Less scrupulous owners would complain so that subsequently, when selling their machines, they could claim that they had been ridden by 'works riders', the implication being that they were 'works' machines. Hence the amazingly large number of allegedly works racing machines in existence around this time. In fact very few complete works machines survived intact. George Hack, being a trained metallurgist, more often than not disposed of racing frames after a relatively short service life rather than run into the hazards of metal fatigue with which he was well acquainted. As a result of this, there are many more original Rudge works engines that have survived to the present day than complete machines.

Rudge-Whitworth also announced its latest sidecars in September 1929, all being mounted on the firm's laminated spring steel chassis. They included: a Streamlined Sports model at £20; the Divan Semi-Sports at £18 18s; the Divan Touring model at £18; and lastly a Light Touring Sidecar at £13 13s. Several tradesmen's versions were also available to special order.

Early in October, three lightweight machines were introduced for the forthcoming (1930) season. They comprised a 248cc ohv and a 248cc side-valve machine similar to the 1929 models and only changed in detail, plus a newcomer — a 300cc side-valve. All three were powered by JAP engines, the ohv being of the latest two-port design which, with its copious Brooklands-type silencers, was a particularly quiet machine for its size. As on the larger Rudge-Whitworth models, twist-grip throttle control was now standardised on these machines as well. The ohv cost £40, while both of the side-valves were priced at £35. On all three models Maglita lighting equipment was £5 5s extra.

For the forthcoming season the dirt-track-racing Rudge-Whitworth retained the former wet-sump motor, but employed a shorter frame and Webb front forks. The end of 1929, in fact, saw something of a Rudge-Whitworth revival on the cinders. One of the leading Rudge exponents at this time was Sid Jackson who, on the Leicester third-of-a-mile (537-m) track, succeeded in October 1929 in setting up a new European lap record at 49.7 mph (80 km/h). This equalled the time and speed put up on similar track by Ginger Lees in Hamburg, Germany.

At the beginning of November the popular Wessex Scramble took place over a course on Salisbury

Plain, for a trophy put up by *Motor Cycling* magazine. The event proved to be a tussle between Vic Brittain (Sunbeam) and Jack Williams (Rudge-Whitworth) — later known as the Cheltenham Flyer — who proved himself the winner with the fastest time of the day to his credit. Run over a very tough course, the event resulted in a mass of broken footrests, shed exhaust pipes and assorted other machine 'débris' being distributed around the course by the end of the day.

Ernie Nott's record-breaking machine was now in such fine fettle that he had little difficulty in persuading John Pugh and George Hack to allow him another crack at the world 500cc hour record, set up at the end of August 1929 by Bill Lacey on his Grindlay-Peerless-JAP at 105.25 mph (169.5 km/h). It was clearly going to be difficult to do this successfully at Brooklands with its now very bumpy track surface, so it was decided to decamp to Montlhéry Track near Paris for the attempt. Thus it was that Ernie made his effort at the French speed bowl on Tuesday, November 19, covering 106.49 miles (171.5 km) in 60 minutes to become the first Rudge rider to hold the record in 17 years. He also took the 100-mile record at an average of 106.45 mph (171.4 km/h). Five days later he made an equally successful assault on the world's 500cc solo 200-mile record at an average speed of 102.9 mph (165.7 km/h).

These records provided excellent publicity material for the Rudge-Whitworth stand at the Olympia Motorcycle Show which soon followed and at which the firm displayed its complete range of models for the forthcoming (1930) season. The orders taken for machines and sidecars at the show, in fact, were better than expected in view of the parlous state of the country's economy in the depression years.

George Hack's research and development department carried out a number of interesting experiments during the year, aimed at machine improvement. For one reason or another, most were not proceeded with, though. One of these involved a novel method of varying an engine's valve timing without removing the timing cover. In this the cam followers were mounted on an eccentric pivot pin, which when rotated caused the followers to move across the top of the cam advancing or retarding the valve timing as desired. A further development of this idea enabled the timing of the inlet and exhaust valves to be varied independently of each other; the exhaust valve's cam follower was mounted on an eccentric spindle which passed through the centre of an eccentric sleeve upon which was mounted the inlet valve's cam follower — there being sufficient clearance between spindle and sleeve to allow their independent rotation.

Despite the high efficiency of the Rudge-Whitworth single-leading-shoe brake, George Hack felt there was room for improvement still and in November 1929 came up with an interesting constant-clearance brake-shoe design. In this there were eight eccentric arms controlling the two brake shoes. Each arm was set at a trailing angle to the brake drum's direction of rotation, each moving in its own slot in the brake shoe concerned. Operation of the brake forced the shoes outwards in such a manner as to provide simultaneous equal contact with the inside circumference of the drum. Although this slightly improved braking efficiency, the unsprung weight of the wheel was unduly increased. It was therefore dropped as an idea for production. As an alternative, an attempt was made to improve the existing braking system by extending the operating arm and cutting serrations on the side of it facing the brake shoe. These serrations were used to provide a form of positive brake operation using a ratchet. Once again the improvement over the standard cam and spring arrangement was marginal and did not justify the additional complication involved and so was not taken further.

A year of successful racing had demonstrated the disadvantages of having only four ratios in the gearbox and serious thought was given by George Hack to increasing this to five. A four-speed gearbox was suitably modified to accept an extra pair of 'cogs', but this resulted in an undesirable increase in its width. The problems this created brought about a rapid cessation of that particular project, which was rather a pity, for a five-speed gearbox would have been a really useful 'first' for Rudge.

The year 1929 ended for Rudge-Whitworth on a note of sporting success. Over in Germany, the year was being brought to a successful close for the marque by the firm's Munich agent and active competitions rider Georg Gschwilm, who, in December, scored a major win at an important ice-race meeting. Nearer to home, both Jack Williams and H.R. Kemble gained first class awards on their 499cc Rudge-Whitworth machines in the London-to-Gloucester Trial, while Miss Betty Lermitte in the same event won the Ladies' Cup riding her 340cc Rudge-Whitworth solo. A great performer in reliability trials of all kinds, Betty used to transport her machine to events inside a covered trailer towed behind her tiny Austin Seven car. She was one of a long line of fine women competition riders who abounded in the late 1920s and early 1930s.

11
TT triumphs (1930)

The year 1930 started well for the company with Joe Sarkis taking his 499cc Rudge-Whitworth to victory, for the second year in succession, in the Senior (600cc solo) Class of the South African TT Races at Port Elizabeth on January 2. He achieved the record average speed of 74.16 mph (119.4 km/h) for this 200-mile (322-km) event. During the course of the race he broke the lap record three times, ending up with an eventual course record of 78.6 mph (126.6 km/h).

With the racing season yet to start in Europe, the first outing of the year for the Rudge-Whitworth racing team was in the Pioneer Run staged on Sunday, February 9 1930. Graham Walker and Ernie Nott both rode 499cc free-engine models, while Tyrell Smith was astride a 1912 Rudge-Multi. Betty Lermitte, who was down to ride a 1913 big-single Rudge in this event, decided not to when Margaret Cottle failed to start and in the process prevented the ladies' team from materialising. Betty, instead, handed her machine over to P.L.B. Wills, who successfully completed the course, as did Walker, Nott and Tyrell Smith.

In the Birmingham Club's Victory Cup Trial a week later, young Jack Williams won the Turner Cup, riding his 499cc Rudge, for the 'best performance by a rider under 21 years of age'. Meanwhile, Arthur Atkinson, the speedway star, was doing very well 'down under' on his dirt-track Rudge machine, during the course of his Australian tour. He arrived in Fremantle on January 10 and the following night cleaned up the Scratch and Challenge Match Races at the local Claremont Speedway and set a new track record into the bargain. At that time Wallis dirt-track specials were being fitted with 499cc four-valve Rudge-Whitworth engines and, riding one of these, Frank Arthur broke a number of Australian speedway records during 1930.

Back in England, Geoff Butcher took top honours in the Cotswold Cup Trial on his 499cc Rudge-Whitworth sidecar outfit, with the best performance of the day. His machine, which was fitted with a special crankcase shield made from steel plate, had a specially modified rear-wheel knock-out bolt and brake anchorage carrying a permanently attached tommy bar as an integral fixture. This enabled the bolt to be held while the nut was removed and then withdrawn, thereby speeding up the process of rear-wheel removal. In addition to the main award, Geoff also collected the Clarke Challenge Cup, the Braham Cup and the Gotham Cup. Quite a field day for him! Young Jack Williams, also on a Rudge, won the Cheltenham Challenge Cup in this trial for the best solo performance and also the Eric Williams Cup.

In March, Herr Georg Gschwilm again made the headlines in the local Munich papers in Germany, this time by winning a mountain race at Garmisch on his Rudge-Whitworth, breaking the previous course record by over 10 seconds. Then, back in England, more honours came to the Rudge marque on the trials' scene with Fred Povey scoring a win in the 250cc solo class of the Leeds £200 Trial.

The year 1930 started off well for Rudge riders with a number of trials' victories.

Rudge FIRST!

Cotswold Cup Trial
March 1st, 1930.
G. R. Butcher.

Cotswold Cup
Best Performance of the Day
G. R. Butcher.

Cheltenham Cup
Best Solo Performance of the Day
J. Williams.

Clarke Cup — G. R. Butcher
Eric Williams Cup — J. Williams
Braham Cup — G. R. Butcher
Gotham Cup — G. R. Butcher

FIVE GOLD MEDALS

Holder of WORLD'S HOUR RECORD

**Rudge-Whitworth Ltd.
Coventry.**

Ernie Nott and Jack Dunfee shortly before participating in their now famous Brooklands match race, in April 1930. Ernie won easily.

Towards the end of the month there also came news of further record-breaking activity in Australia. At Melbourne, in Victoria, a 499cc Rudge in the hands of Jimmy Wassal covered a flying-start ¼-mile (402.5 m) at a mean speed of 109.75 mph (176.73 km/h) to set a new 500cc solo Australian national record. Rueb Wheeler, riding the same bike, averaged 107.14 mph (172.53 km/h). Alan Bruce drove it over the same distance with a sidecar attached complete with passenger and set a new Australian national record for 500cc sidecars at a mean speed of 95.79 (154.25 km/h). All the timing was carried out electrically.

Back in England, thoughts had been given to the formation of a Rudge Enthusiasts' Club (a forerunner of the present club). With this in mind, J. Huxam, the firm's Bournemouth agent, staged a rally for Rudge owners on Sunday, March 16 1930, with Bournemouth as the venue and the Club was launched by popular vote of all those present under the patronage of Graham Walker who was present with his wife.

On April 5, the British Motor Cycle Racing Club (BMCRC) celebrated its 21st anniversary with a specially organised race meeting at Brooklands Track and Ernie Nott took part on his 1929 works racing machine, the 1930 works racers not yet being ready. After a series of short handicaps and one-lap sprints, Ernie came to the starting line on his 499cc Rudge-Whitworth for a three-lap match race set up as a promotional stunt by Graham Walker. This was to be against the well-known car racer Jack Dunfee and his 2976cc Ballot.

The roar of the Ballot sounded in strange contrast to the crackle of the Rudge as a rolling start was made, a big touring car acting as the pace maker. Nott beat the starting flag slightly, while Dunfee lost some time in getting his car really moving. As a result the motorcycle had a useful lead of some 100 yards (91.5 m) or so by the time that the Members' Banking had been reached. On the Railway Straight Jack Dunfee closed up, but he still could not catch the flying Rudge. Nott managed to keep just safely ahead until the finish, winning by a matter of 30 yards (27.4 m) at 99.61 mph (160.40 km/h). Both Nott and Dunfee did an extra lap to make sure of it, and both had an amusing word or two to say about the race afterwards over the loudspeakers. Dunfee's young female passenger, an attractive brunette, was having her first taste of Brooklands, and said she found it 'exhilarating but bumpy!' Another prominent Rudge rider at this meeting was Gus Grose, who competed in several match races.

Geoff Butcher, as usual, rode his Rudge outfit with distinction in the Kickham Trial, winning the Dickinson Cup, the Bristol Cup and the Bath & Avon Rose Bowl. Of the other Rudge riders taking part, five gained first-class awards, three got second-class awards and five others, including Betty Lermitte, won third-class awards. In the Scottish Six Days' Trial in May, Betty did even better, winning a Silver Cup riding her 350 Rudge. Eleven Rudges were entered in this event and only one lost a point, gaining six Silver Cups and five Gold Medals in the process. The Rudge-Whitworth solo works team that took part in the 'Scottish' comprised Jack Williams, Fred Povey and Bob MacGregor. Unfortunately, when in the process of winning the Manufacturers' Team Prize for over 350cc machines, the trio's chances of doing so were permanently blighted by MacGregor's machine breaking its chain on the very last hill of this arduous six-day event. This was compensated for to some extent by MacGregor winning the Rake Hawe Cup in the Alan Trophy Trial, together with a Gold Medal. Golds were also awarded to Fred Povey, C.R. Sanderson and Geoff Butcher (sidecar).

Meanwhile, down in South Africa, Joe Sarkis, riding from scratch on the Rudge with which he won the South African TT Race, was the winner in the 75-mile (120.8-km) handicap race at the Natal Spruit Race Meeting.

By the end of April 1930, the new works racing models were ready and the Rudge team took them off to Ireland for the North-West '200' Road Race. Staged in May, it seemed a heaven-sent opportunity for testing the new racers prior to the TT Races in the Isle of Man. Although basically the same as the 1928 and 1929 machines, which had so very nearly won their respective 'Island' races, the new 499cc engines had longer connecting rods to help reduce side thrust on the pistons and longer induction tracts to help improve acceleration. The modifications necessitated slightly raising and lengthening the frames. It was questioned whether they would steer well and whether their 29 to 30 bhp at 5600 rpm would be sufficient to beat the opposition? The answers to both these queries seemed to be yes. For 15 of the 18 laps of the race Graham Walker led, until a broken oil pipe put him out of the running, whereupon Ernie Nott took over the lead and went on to win at a record average speed. Stanley Woods on his much-modified camshaft Norton, could do no better than third behind a Rudge 'privateer' named Patrick Walls. The auguries, therefore, seemed good for the Isle of Man Senior TT Race. The new frame steered better, if anything, than the 1929 version and the oil pipe breakage was a minor irritant easily remedied. All this lifted a lot of worry from the shoulders of George Hack, who at that time was up to his eyeballs in work on a completely new project, a 350cc ohv engine with four radially disposed valves. Clearly anything seriously amiss with the Senior TT machines, at this late stage, would have been disastrous.

While all this was going on, news came through that a Continental rider named Steinfellner had taken his Rudge to victory in the 500cc Class of the Austrian TT Race at record-breaking speed, some six minutes ahead of the second finisher.

The original intention regarding the new 350 radially-valved Rudge was for Tyrell Smith to ride a machine fitted with the new engine in the Junior TT as an experiment. Since, in the past, Rudge-Whitworth had had no pretensions to Junior TT honours, any failure here would not have such deleterious effects on company sales as it would have done in the Senior Race if the engine had been a 500cc motor. The situation was to change though.

The new engine had a bore and stroke of 70×90 mm, but unlike previous Rudge designs possessed a hemispherical combustion bowl. The radial disposition of the valves complicated the valve-operating gear arrangement, but an ingenious and neat solution to the question of how to operate them satisfactorily was achieved by using six overhead rockers. Each pushrod was associated with three rockers, the first lying parallel with the longitudinal axis of the frame and the other two parallel with the transverse axis of the frame. The underside of the end of the second rocker lay beneath the inner end of the third rocker and caused it to depress the nearside valve. An important feature was that the geometry of the layout was such that each of the rockers had true rolling motion at its points of contact; it was claimed that no wear-inducing rubbing existed at all.

Tyrell Smith's arrival in the Isle of Man TT paddock for the first morning's practice with the new 350 caused more than a few raised eyebrows. The cynics were asking: 'will it work?' The question was soon answered with a vengeance, when Tyrell's standing-start lap of 31 min 29 sec (equivalent to 71.9 mph or 115.78 km/h) knocked 26 sec off Freddie Hicks' Junior TT lap record! 'Ah, but this is practice. Would it stick it in the race?' was the cynics' rejoinder. The answer to that question was unknown at that stage, even to George Hack — but he was praying hard! True, a prototype engine had performed remarkably well on the test bench, despite being flogged unmercifully for hour after hour. The three 350 racing engines available, however, had only had a brief run each on the test bench back at the Coventry works. Then, just to make life really difficult for George Hack, John Pugh, his Managing Director, decided that Ernie Nott and Graham Walker should also ride, not without considerable protest from George. He took a very dim view, in particular, of Graham parking his 15-stone (95.45-kg) weight on top of a little

Trials' rider Betty Lermitte won a silver cup on her 350 Rudge in the 1930 Scottish Six Days' Trial.

Rudge
REG. TRADE MARK.

Holder WORLD'S HOUR RECORD

Miss Betty Lermitte—Rudge 350—won a Silver Cup.

FIRST!
100%
Reliability
in
Scottish Six Days' Trial.

11 Rudges entered winning

6 Silver Cups
5 Gold Medals

Not a single mark lost on Time or Condition.
The most convincing proof of Thorough Dependability Ever Made.

HAVE YOU JOINED
The Rudge Club YET?

Rudge-Whitworth Limited, Coventry.

Coupon for Free Booklet and Particulars of Hire Purchase Terms.

untested 350 experimental model and especially so when he saw all his precious 350 spares being depleted to build two extra machines. There was, therefore, an anxious silence as the cylinder head and cylinder barrel of Tyrell's machine were removed after his sensational morning-practice debut. At first all seemed to be well, then a cloud of gloom settled over the Rudge teamsters, when it was discovered that both piston bosses were cracked! In view of the shortage of spares and of time available before the race, desperate decisions were called for so the piston was refitted 'cracks and all', and matters left in the laps of the gods!

Below *The record-breaking 1930 TT Rudges had this type of cylinder head with fully-radial exhaust and inlet valves.*
Bottom *Parallel inlet and parallel exhaust valves were fitted on the 1930 Senior Rudge engines.*

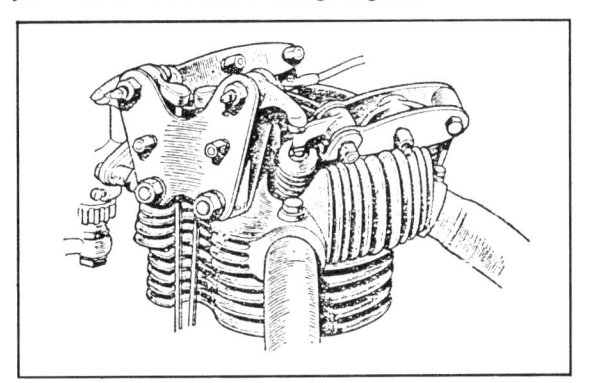

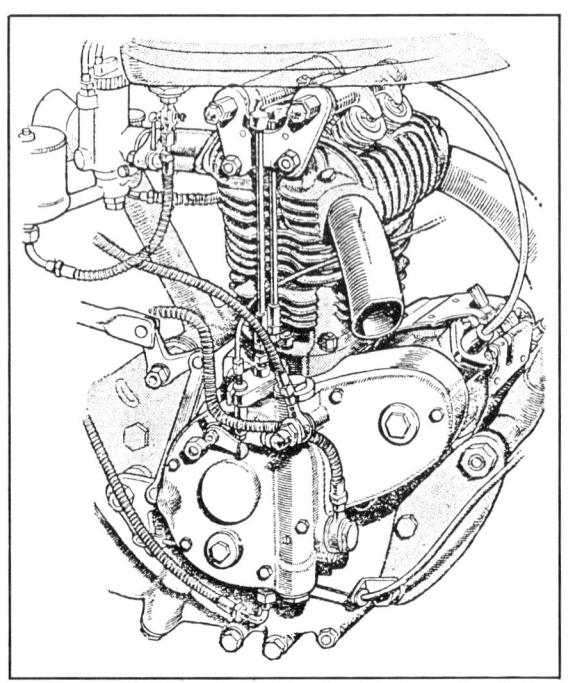

The next day Harold Willis (Velocette), Jimmy Guthrie (AJS), Charlie Dodson (Sunbeam) and Freddie Hicks (AJS) jointly headed the Junior practice leader board with 32-min laps apiece: the clear inference being that the new Rudge just could not stand the pace. The answer to this came during the third morning's practice period, when Tyrell Smith completed three consecutive laps, the last in 31 min 54 sec or at 70.96 mph (114.27 km/h) and, much to George Hack's relief, the piston cracks had not spread.

While all this was going on, Ernie Nott and Graham Walker were busy putting in some practice on their 500s for the Senior event, and it was not until the second week of practising that they made an appearance on their Junior Race mounts. On his first practice lap Graham Walker felt his 350 starting to seize at the tip of Craig Willies Hill. The following morning his engine started to seize at precisely the same spot on the course again. Matters were now getting serious, for the engines of both Ernie Nott and Tyrell Smith had also acquired cracks in their piston bosses! Graham set to with fine files and relieved the high spots on his seizure-prone piston and completed his remaining qualifying laps, but seized yet again during the final practice session on the Saturday morning. This time it resulted in a ruined barrel and piston and, worst of all, a bent connecting rod.

Since Graham Walker's machine had to be handed in for the race at 5:30 pm that day, there was nothing for it but to rebuild its engine from George Hack's precious prototype. Then, fortified with a Guinness inside him, he took this reassembled machine out to Kirkmichael and back just to let everything bed down. Before reaching Ballacraine, though, the 'dragging sensation' made its presence felt again. A change of sparking plug and a new magneto made no difference, either. It was all very puzzling to Graham. Then, later, walking back to the hotel with George Hack, the 'penny dropped'. It must be the one item not checked that was the culprit, namely the gearbox. There was a possibility that its plain sleeve bearing was on the point of seizing and in so doing putting an unduly high load on the engine. The only thing to do was to check the gearbox out, but the machine had been handed in to the race officials. On the Sunday morning, therefore, the Rudge teamsters made their appeal to open up Graham Walker's gearbox on the grounds of rider safety. In the presence of ACU officials the box was stripped and George Hack removed the apparently offending item. It was in good order inside though, so it was decided to reassemble and to allow the machine to start. Then came disaster, when Jock, the Scots mechanic, cried: 'Ye've drappit in a roller, Mr Hack'. Nothing could be done though at that late stage and sadly the machine had to be returned to the official tent with a special prayer for the morrow.

That evening at the pre-race conference, the grim-faced Rudge teamsters were given their riding instructions for the race. Graham Walker was told to only use half throttle until he had passed Craig Willies and only three-quarter throttle for the rest of the lap. If after that the roller hadn't jammed in the gear teeth, Graham could go as fast as he liked. In later years, Graham recalled that: 'I did not sleep too well that night!'

The final tactics were that Tyrell Smith was to set the pace on the first lap with Ernie Nott backing him at a slightly slower rate. It was obvious to Graham Walker, in view of his instructions, that George Hack did not expect him to finish. After a brief homily from the latter about the necessity of keeping their engines at all times below 6400 rpm, with mixed feelings, the Rudge trio made its way to bed.

The morning of the Junior Race dawned brilliantly fine. The original 46 entries were reduced to 43 starters and at 10:00 am sharp the first man, Tim Hunt (Norton) was sent on his way. Then 2½ minutes later Graham Walker was sent off on his half-throttle tour to Craig Willies. He passed the fatal seizing point without incident which, as he later recounted: 'lifted a ton weight of worry off my back'. He advanced the throttle to the three-quarter open position as per pre-race instructions and completed his standing-start lap with his engine still in one piece lying 14th.

Charlie Dodson (Sunbeam) led at the end of the first lap, averaging 71.09 mph (114.48 km/h) with Tyrell Smith only 3 sec slower at 70.99 mph (114.32 km/h). On Lap 2, Tyrell and Charlie became joint leaders, averaging 71.28 mph (114.78 km/h). Meanwhile Jimmy Guthrie (AJS) had moved up from sixth to fifth spot taking the lap record at 71.61 mph (115.31 km/h) in the process and Ernie Nott had moved into sixth berth at an average of 70.52 mph (113.56 km/h).

The third circuit saw Dodson drop out with valve trouble, letting Tyrell into first place with a new lap record of 71.69 mph (115.44 km/h). Then, at the end of this lap, when filling up at the pits, he heard the announcement over the loudspeakers that Jimmy Guthrie had cracked this with a speed of 71.96 mph (115.9 km/h). Meanwhile Graham Walker, now running on full throttle, had moved up into ninth place. The crowd was now getting really excited.

At the end of the fourth round, Tyrell Smith had increased his lead over Guthrie to 28 seconds, having averaged 71.09 mph (114.5 km/h). Ernie Nott had now moved into fifth spot at 70.49 mph (113.51 km/h), with Graham Walker sixth at 70.46 mph (113.46 km/h). Three prototype machines on the leaderboard! It was unheard of! Excitement was now running at fever pitch amongst the crowds watching the race! But compared with this the fifth

Here is a tired, happy Ernie Nott after finishing second in the 1930 Junior TT race on one of the new radial-valved Rudges that swept the board that year. Standing behind the machine to his right is John Pugh, the Rudge-Whitworth Managing Director.

circuit was a real heart stopper! First a rider called Barwell crashed at Quarter Bridge breaking a couple of limbs. Then Freddie Hicks stepped off his AJS at the same spot, reboarding his machine again only to retire at Kirkmichael with an engine fault caused by the crash. Then Willis retired with engine trouble on his Velocette. The result of all this was that at the end of Lap 5 the Rudge teamsters occupied three of the first four places, Jimmy Guthrie still holding on to second spot on his AJS. Then, on the penultimate lap, Jimmy retired with a broken rocker. This let the three Rudge teamsters into the first three positions, which they maintained to the end of the race.

At the finish, Tyrell Smith flashed over the line first to average 71.08 mph (114.44 km/h) for the seven-lap event — the first ever Junior TT win by a Rudge, let alone an untested prototype. It was also the first Junior TT Race to be won at over 70 mph (112.7 km/h) and the last to be won by a pushrod-engined machine.

Ernie Nott came home second at 70.89 mph (114.15 km/h) and also, incidentally, set up a record lap during the race of 72.22 mph (116.30 km/h). Graham Walker made it a 'hat trick' with his third place at 70.77 mph (113.96 km/h) and enabled Rudge-Whitworth Ltd to take the Manufacturers' Team Prize as well. His performance, in view of his machine problems and weight disadvantage plus his war wound, was probably the most remarkable of all.

When subsequently stripped for examination and measurement, the only faults the very experimental power units used by the Rudge team showed consisted of a broken, but still functioning, inner inlet valve spring on Tyrell Smith's motor and an inner exhaust spring in a similar condition on Ernie Nott's engine. The Rudge teamsters derived the greatest of pleasure in embarrassing their race rivals by drawing the official machine examiners' attention to the cracks in their piston bosses. The biggest joke of all, however, came later back at the Rudge camp when, stripping down the gearbox of Graham Walker's machine, it was revealed that *all the rollers were in place.* He had had hours of anticipatory misery for nothing! As to the cause of the mysterious 'dragging sensation', George Hack attributed it to 'rider temperament'. So ended an amazing Junior TT Race by any standards and, as John Pugh put it at a private celebration afterwards, the evening before the official prize presentation ceremony: 'They were entered purely as an experiment, gentlemen. The experiment has proved successful!'

The Rudge teamsters now switched their attention to the Senior event, scheduled for Friday, June 21. For the 1930 TT races practising period George Hack had organised a system to cope with the fact that it had been shortened from 12 to only nine days. This entailed Graham Walker concentrating on the 500s, Tyrell Smith on the 350s and Ernie Nott sharing his time roughly equally between both.

The first morning practice on the Thursday put Stanley Woods (Norton) at the top of the leaderboard in the Senior category, with Graham Walker in third spot behind C.J. Williams on the Raleigh. Meanwhile Wal Handley, whose FN machine seemed unlikely to be ready in time for the race, was furiously trying to get ACU permission to change the make of his race machine. His hopes of a mount in the Senior Race lay in someone standing down in

Cartoonist Collis' view of the 1930 Senior TT Race winner — Wal Handley. The machine is clearly not a Rudge, the drawing being made before he joined the team.

his favour. With this in mind, George Hack got together with Wal and Jim Whalley, who had offered to nominate Wal as the rider of his privately entered 499cc Rudge-Whitworth. John Pugh dearly wanted Wal to ride a works' machine, but felt that it was unfair on Tyrell, Ernie and Graham, to impose it on them and so left it up to them to come to a decision about it amongst themselves. Much to their credit they unstintingly agreed. Wal Handley was popular as a rider, brilliant in his own way but with a fiery temper and a strong will of his own. It was this last aspect of his character that George Hack was clearly meaning to master from the outset.

It was an amazing morning. After the decision had been made, George Hack called Wal Handley and Jim Whalley into the room where the other riders were. Obviously delighted, Wal's reaction to the news was brief and to the point. 'Good!' he said. Then, as they all walked out to the garage where the machines were prepared, George, as Team Manager, made one of his bluntest ever speeches. Having heard that Wal could be 'difficult' at times, he made it abundantly clear that what he (George) said 'went' at all times so far as racing in the Rudge-Whitworth team was concerned. To everyone's amazement, Wal took all this without turning a hair. From then onwards an air of mutual respect of each other's abilities evolved between the two men, which provided a very harmonious working relationship. George Hack, with a stroke of genius, had adopted the right approach to Wal Handley and during the whole of the practising period he became the perfect team man with no sign of temperament whatsoever.

Wal responded to the friendly working atmosphere of the Rudge camp and soon knuckled down to the serious business of the week. In the second morning's practice period, he seized up the gearbox of his hastily prepared Rudge, but with this freed and repaired he made second fastest Senior time on the third morning in 32 min 3 sec with Graham Walker third in 32 min 15 sec. It was not until the Friday, however, that Wal really showed the stuff from which he was made, when he went round in 30 min 7 sec at 75.20 mph (121.10 km/h), knocking some 40 seconds of the Senior TT lap record.

Before the final handing in of the machines, for locking away before the race, Handley's machine was given the final 'once over', with George Hack's permission, by Sammy James, Wal's mechanic and inseparable friend. It was hardly necessary, though, after Jim Dalton and 'Goldflake' Wills, the official Rudge mechanics, had finished their handiwork. Meanwhile, brimful of confidence about his team's

An impression of Wal Handley at Craig-ny-Baa during the 1930 Senior TT Race, by artist Roland Davis.

chances, John Pugh had tried to back them with a local bookmaker for a 1-2-3-4 finishing order in the Senior, but the bookie could not stand the odds and returned his money, which was fortunate for him as it turned out. But, for real 'one-upmanship' in the war of nerves against rival camps, Wal Handley put £20 on himself with the local bookie to lap the course in 29 min 40 sec — 67 seconds faster than the 1929 lap record! The main worry, though, was whether the weather, which looked decidedly threatening, would hold. Then, on Thursday evening, the storm burst and rain lashed down on the course. The Rudge teamsters went to bed decidedly unhappy, visualising their ride the following day being dominated by water in magnetos, water on the sparking plugs, etc. In the morning, though, there was a glorious sunrise and as the starting time approached the roads had dried out, except under the trees.

By the time the race had started, Wal Handley was the favourite to win. The first lap was a sensation. He covered it in 29 min 47 sec from a standing start, breaking the 1929 record by a minute at a speed of 76.03 mph (121.43 km/h). Tyrell Smith was through second in 30 min 21 sec at 74.60 mph (120.12 km/h), with Graham Walker eleventh and Ernie Nott well back.

On Lap 2, Wal and Tyrell maintained their race positions, Graham had moved up to seventh spot and Ernie Nott now lay twelfth. On the third lap,

Wal came in to refuel, yet, despite this, completed the circuit in 29 min 41 sec at 76.28 mph (122.83 km/h), which turned out to be the fastest lap of the race — Rudges now occupying the first three race positions, Tyrell Smith being second at 75.00 mph (120.77 km/h) and Graham Walker in third spot at 74.15 mph (119.40 km/h). Ernie Nott had moved up to ninth position.

Rain on Lap 4 slowed the pace down slightly, but the first three positions remained the same. Meanwhile Ernie Nott, on the fourth Rudge, had moved up to eighth spot. On Lap 5, Tyrell Smith, whose Rudge had lost power and was misfiring, dropped back to fourth position. This, at the time, he thought was due to the experimental 14-mm sparking plug he was using, but subsequently he discovered that only one of his two exhaust valves was functioning due to the fracture of a rocker finger. Meanwhile Ernie Nott had moved up to sixth position, averaging 72.50 mph (117.23 km/h).

On the sixth and penultimate lap the rain really started to pelt down; riding leathers were becoming sodden and race conditions degenerating rapidly. Amazingly, race speeds had not suffered very much from the appalling weather, Wal Handley still being out in front, averaging 75.58 mph (121.71 km/h).

Taking no chances, Wal completed his final lap in a safe 33 min 45 sec coming home to a brilliant victory in ghastly riding conditions, at an average speed for the seven laps of 74.74 mph (120.35 km/h). Graham Walker was second at 73.10 mph (117.71 km/h), Tyrell Smith sixth at 71.11 mph (114.51 km/h) and Ernie Nott, seventh. The only other Rudge rider in the race, the New Zealander, Percy Coleman, retired. For their fine efforts the three 'official' Rudge teamsters, Walker, Smith and Nott, won for their company the much coveted Manufacturers' Team Prize.

The incredible thing was that Wal Handley's *average* speed for the race was actually higher than the 1929 lap record, despite the weather conditions. It is therefore interesting to note that, although he undoubtedly had the fastest of the works Rudges at his disposal, which would otherwise have been ridden by Tyrell Smith, it actually used a lower compression ratio than the other works machines, 7.0:1 as opposed to 7.25:1.

In this context it is worth noting some of the other technical details of Wal Handley's historic machine. The lower compression ratio was achieved using two compression plates under the base of the cylinder. The aluminium alloy piston was fitted with two 1.5-mm deep Brico rings, the top one having a gap clearance of 0.018 to 0.022 in (0.46 to 0.56 mm) and the bottom one a clearance of 0.010 to 0.014 in (0.25 to 0.36 mm). Carburation was by means of a Type TTA 29 twin-float Amal instrument, fed by all-Flexicas, flexible petrol piping. It pulled a larger main jet than the other Rudge machines in the works team, a No 62, and had a bore of $1\,5/32$ in (29.37 mm). It was mounted on a 1½-in (38-mm) long extension piece to give a longer induction tract. The engine employed straight exhaust pipes terminating just beyond the rear wheel hub and no megaphones were fitted. The sparking plug used for the race was a KLG 348. Other details included the fitting, by Wal's choice, of a 27 × 3 in (178 × 76 mm) ribbed tyre on the front wheel and a 27 × 3.25 in (178 × 83 mm) studded tyre on the back. The eventual gear ratios he used were: 4.4, 4.8, 6.3 and 8.75:1.

The machine was kept by Wal Handley until the end of the 1932 season and then sold to Reg Wood. The latter used it in both solo and sidecar form at the pre-World War 2 Leicester Super Speedway, which was similar to any other speedway but longer.

The Rudge-Whitworth advertisement following Wal Handley's remarkable win in the 1930 Senior TT Race in the Isle of Man.

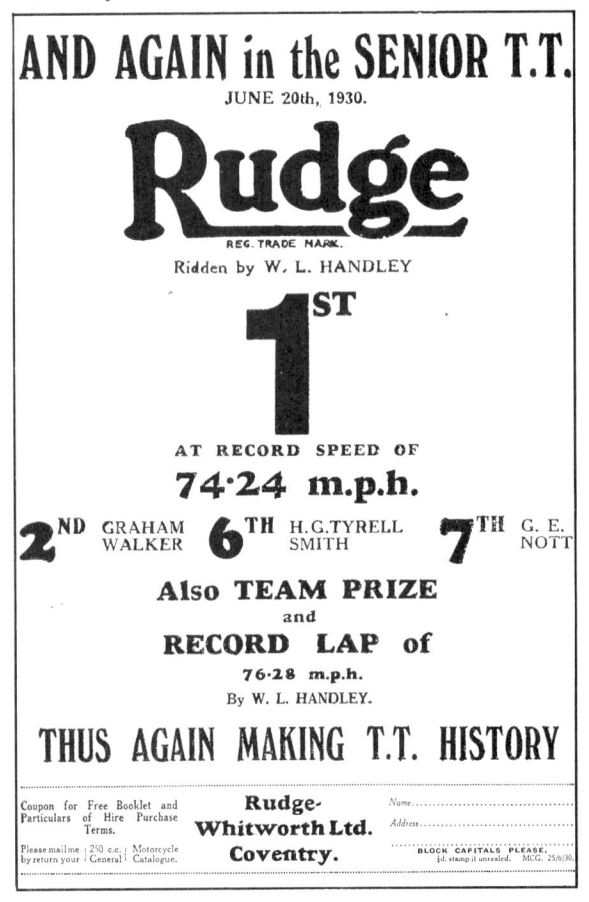

AND AGAIN in the SENIOR T.T.
JUNE 20th, 1930.

Rudge
REG. TRADE MARK.

Ridden by W. L. HANDLEY

1ST

AT RECORD SPEED OF

74·24 m.p.h.

2ND GRAHAM WALKER   6TH H.G.TYRELL SMITH   7TH G. E. NOTT

Also **TEAM PRIZE**
and
RECORD LAP of
76·28 m.p.h.
By W. L. HANDLEY.

THUS AGAIN MAKING T.T. HISTORY

Coupon for Free Booklet and Particulars of Hire Purchase Terms.
Please mail me 250 c.c. Motorcycle
by return your General Catalogue.

**Rudge-Whitworth Ltd.
Coventry.**

Name............
Address............
BLOCK CAPITALS PLEASE.
Jd. stamp if unsealed.   MCG. 25/6/30.

After that war, the Rudge was raced in the sidecar class at Cadwell Park.

With all the successes in the Junior and Senior TT Races, the fact that George Himing had ridden a 245cc JAP-engined Rudge into eleventh place in the Lightweight event had passed by almost without notice. All in all, Rudge-Whitworth Ltd had a very successful 1930 TT race week. Little could John Pugh or George Hack have realised that it was to be the last occasion on which Rudge-Whitworth machines would win the Senior and Junior Tourist Trophies.

The piston-boss-cracking and piston-seizure troubles experienced in 'The Island' now initiated a series of experiments to try and overcome these problems before the next big event took place. Fully skirted pistons were found to seize after only 5 to 10 min of full-throttle running on the test bench, while the Invar-strutted type lasted, on average, only about 18 min. When the skirt was relieved above and below the gudgeon pin bosses, the pistons survived longer, but the cracking problem then started to rear its ugly head again. The result of all this work was the designing of a new 'slipper' type piston, which would run for an hour with the engine flat out on the firm's Heenan & Froude brake, without any signs of either cracks or seizure developing. The problem was solved! This piston, therefore, was standardised on both the works racers and production sports machines.

The Rudge racing team, after its great success in the Isle of Man, might reasonably have expected to have had something of a 'cake walk' in the various European races in which it entered: but this was not to be. Unlike English events, there was a tendency in Continental racing to run off all the different capacity races together on the same day. This prevented a rider from running in more than one class and so it was not possible for Rudge-Whitworth Ltd to field a full team of riders in both the 500cc and 350cc classes. Since the new 350 radial-valved Rudge was still in the development stage — despite its Junior TT success — it was decided to concentrate on the 500cc races. Graham Walker and Tyrell Smith, therefore, were delegated to ride 500s and Ernie Nott the 350 in most of the foreign events entered by the firm.

The strategy proved successful in the 500cc races, but met with only two successes in the 350cc class. At the first major European race meeting after the Isle of Man TT Races, the German Grand Prix held on June 29 at the Nürburgring, Graham Walker came home first in the 500cc event with Tyrell Smith third behind Stanley Woods and his Norton. Ernie Nott's one-and-only 350cc class win on the new radial-valve machine, on the other hand, was in the European Grand Prix, staged at Spa in Belgium. He came home some 8 minutes ahead of Arthur Simcock (Sunbeam) at an average speed of 70.60 mph (113.78 km/h). In the 500cc class in the same Grand Prix, Tyrell Smith came back into winning form, averaging 74.26 mph (119.58 km/h) and set up a new course record at 76.00 mph (122.38 km/h) for the lap, with Graham Walker second.

In 1930 Rudge dominated dirt-track racing and, as Sales Manager, it was Graham Walker's job to do what he could to propagate sales of Rudge dirt-track machines. One day five Rudge speedway machines belonging to the Brandon Track were burned out in a lorry crash and, as the riders had no cash with which to buy replacements, Wilmot Evans staged a charity race meeting at which Wal Handley, Tyrell Smith, Ernie Nott and Graham Walker promised to make fools of themselves, for the benefit of the moneyed public. In the event, Jack Parker substituted for Wal, and they all mounted enormous bare-backed cart horses with the idea of covering a few laps. As Graham recounted some years later: 'We were started by some maniac with a 12-bore shotgun and my nag, more from fright than anything else, jumped into the lead with me clinging around its neck for dear life. This was necessary because its back was so broad that even my long legs stuck out nearly at right angles! At this point, that wily and experienced dirt-tracker Jack Parker cut inside me, his horses's shoulder catching my left boot and hurling me off.' The result of all this buffoonery was irreparable damage to Graham's right ankle which, added to the limitations imposed by his war-wounded left leg, meant that from then, until he retired from active competition riding in 1934, he had to rely largely on his arms and obstinacy for riding control.

Graham Walker was in severe pain as a result of this incident and George Hack decided that the Ulster Grand Prix would determine whether he could go on riding as a member of the Rudge-Whitworth works racing team or not. Meanwhile, back at the works, George had another pet project up his sleeve in his efforts to extract more power from the Rudge design. This was a three-valve cylinder head, with one inlet and two exhaust valves. Designed in August 1930, this reduced the reciprocating weight on the inlet side, but it also cut the total available inlet valve area, with the result that engine filling was not quite so good as that of the four-valve layout. The net result was a poorer performance than that of the four-valve design so the project was stopped.

The team arrangement used in the Ulster Grand

Prix was the same as that used for the major European races, with 'Nottie' riding his radial-valve machine in the 350cc class and Tyrell Smith and Graham Walker on the 500s in the 500cc event. In contrast to the previous year, the weather for the 1930 Ulster was awful, heavy rain falling intermittently all day. In the 350cc race, Ernie Nott took the lead from the start and held it until Lap 3 when, with a determined effort, Freddie Hicks (AJS) grabbed front spot. This, Hicks held until Lap 8, when Ernie once again took over. On Lap 9, Leo Davenport (AJS), who now lay second, put the pressure on 'Nottie' and finished the race 4 minutes ahead of him to win. Ernie came home a good second at 73.71 mph (118.70 km/h).

At the end of the first lap in the 500cc race, Stanley Woods (Norton) and Graham Walker, who was given starting assistance in view of the recent injury to his left ankle, were neck-and-neck with Charlie Dodson (Sunbeam) only a few yards astern. On Lap 2, Stanley dropped back and it was Charlie now who diced for the lead with Graham. Then, on the third lap, the prospect of Graham Walker winning the Ulster for the third year in succession was wiped out, when he was forced to retire with water in the magneto. Meanwhile, Tyrell Smith had moved up to third spot and that was how he finished, averaging 79.57 mph (128.13 km/h).

To complete a relatively successful year of road racing for Rudge-Whitworth Ltd, Ralph Merrill, a 23-year-old wine merchant from Didsbury, scored an outstanding win in the Senior Manx Grand Prix in September. G.W. Wood, another Rudge rider, led Merrill at the end of the first lap, but on the second Merrill overtook Wood and thereafter never relinquished his lead. Merrill's winning average was 69.49 mph (111.90 km/h) and he also made the fastest lap (his third) at 71.13 mph (114.54 km/h). Wood, a 25-year-old building contractor from Liverpool, finished second at 69.38 mph (111.72 km/h) and Norman Gledhill (Norton) third at 68.61 mph (110.48 km/h).

Unfortunately this fine victory engendered a certain amount of bad feeling and there were insinuations made to the effect that the two Rudge riders were mounted on genuine works machines and not models purchasable by the general public, as required by the race regulations. As a result, John Pugh went to great lengths in the press and elsewhere to show that the real works racing machines were being raced at a totally different venue at that time by factory teamsters. To some extent it was inevitable that such a furore would be caused when the road-going sports machine and the private owners' racing machine were both listed in the

John Pugh had something to 'crow' about at the end of 1930, as this Rudge agent's advertisement shows — but the Manx Grand Prix victory backfired somewhat and led to a tidying up of its race entry requirements.

catalogue under the name 'Ulster', which was confusing enough for those who were Rudge enthusiasts, let alone those who were not. Anyway, as a result of all this and the fact, also confusing to the non-Rudge man, that the firm's production year started in mid-season, the private owners' racing machine became known as the Rudge-Replica.

It was shortly after the Manx Grand Prix, in October 1930, that Rudge-Whitworth investigated an interesting modification to the radial four-valve operating gear, suggested by a Mr Watkinson. The idea was that only four rockers would be needed to operate the four valves, instead of the six then used, if the right-hand rockers ran at an angle of 45° to the pushrod, across the top of the valves. It was suggested that, where the rockers met, the circumference should be gear cut so that they meshed. George Hack modified this design so that the remote rocker was operated by the normal Rudge leverage technique. Unfortunately, it would have entailed eight bearing pillars in the head, which put it right out of court commercially.

The month before, Ernie Nott, never willing to miss the chance of a Brooklands outing, took his machine down to Weybridge for the Brooklands' 200 Mile Races, postponed from September 13 when they were washed out by rain. The day of the races proved to be sunny, not too hot and with little wind, ideal in fact for high speeds. At the end of the first lap, Ernie Nott led, closely followed by Bert Denly (AJS). The other two Rudge riders in the race, Gus Grose and Jack Levene, lay fifth and last (eighth) at this stage of the game. On Lap 5, Denly passed Ernie and at the end of quarter distance (18 laps) still occupied first spot, Nott having averaged 101.3 mph (163.12 km/h) in second place. Gus Grose now lay fourth and Jack Levene fifth.

At half distance (36 laps), Denly still led, and Ernie Nott was still second, at an average speed of 102.16 mph (164.51 km/h) only marginally under his own 500cc solo 100-mile record average. Grose lay third (a lap behind) at 98.14 mph (158.03 km/h) and Jack Levene (close behind Grose) in fifth place.

Two laps later, Ernie came in to refill with fuel and oil, losing half a lap in the process. His engine started to misfire, which was not a good sign. By Lap 49, the misfire was really bad and he came in and changed his sparking plug. This seemed to solve the problem, but he had lost a couple of laps to the race leader, Denly, by this time. Meanwhile, on the same lap Grose pushed *his* Rudge in and retired.

At the three-quarter distance (54 laps), the leaders were Denly (AJS), 100.82 mph (162.33 km/h); Nott, 96.75 mph (155.84 km/h); and Jack Levene, 95.21 mph (153.32 km/h). On Nott's 58th

The 1931 Rudge 'Ulster' on show at the Olympia Motorcycle Show in November 1930.

lap, disaster struck — a length of tread flew off the back tyre and struck his rear end with a resounding thwack. He retired and a commiserating Rex Judd commented: 'It must have reminded him of his school days!' Ernie was not amused!

The race finished with Bert Denly the winner at 97.26 mph (156.62 km/h). Jack Levene came home a worthy second on his Rudge at an average speed of 96.1 mph (154.75 km/h). When Denly crossed the finishing line, Jack was less than a lap behind. After finishing, the latter went on to do another lap in order to reach The Fork again (where the race started); during this lap, he, like Ernie Nott, also had the tread come off his rear tyre and it burst, so that he had to push in from the far side of the track.

The year finished as usual with the Motorcycle Show in November, the premier position on the Rudge-Whitworth stand being taken by the new radially-valved 350s. The 350cc and 500cc Replicas were now available in place of the tuned Ulsters. Two hush-hush projects were also underway.

The 245cc and 300cc JAP-engined models had been dropped since their engines had to be purchased from J.A. Prestwich of Tottenham, and a new home-bred 250 engine was under development as a replacement power unit. The second scheme was slightly more exotic. This consisted of the design and development of a vee-four engine intended for an attempt at the World Motorcycle Speed Record. This was very much John Pugh's baby and very little information exists as to the details of this project. He approached Captain Irving, famous as the driver of Sunbeam record-breaking cars at that period, to assist with the design of a suitable streamlining. Unfortunately, the latter had so much work on of his own at that time that he had to decline and so the project was stillborn.

12
Hard times (1931-1932)

By 1931, the world-wide economic depression had started to bite really hard. The motorcycle industry was beginning to feel the pinch, and the sales of new machines were beginning to falter. Some 7000 Rudge-Whitworth motorcycles were sold during 1930, the same as in 1929, but the end of 1931 saw this drop to an alarmingly low 2500. Hard times were here in earnest for everyone.

A complete redesign of the whole Rudge range of machines was under way by December 1930, in an attempt to find a formula that would improve the sales position, and in the New Year the first result of this emerged in the form of six new models plus a Streamlined Sports Sidecar selling at £22. The '350 Standard' and '350 TT Replica' models, selling at £54 10s and £82 respectively, each had four radially-disposed overhead valves like the 1930 works' 350cc racing machines. The 500s were available in four different forms: the Standard Ulster, at £70; the Ulster Replica, at £85; the Special road machine, at £59; and the Dirt Track machine priced at £79. The TT Replicas, as the name implies, were exact copies of the 1930 works racers. All of the 500s retained the parallel-valve pent heads used in previous years.

A major design departure on all of the 1931 models was the placing of the magneto behind the engine. Despite the wholesale success of Rudge-Whitworth machines in competitions during 1930, the firm's Sales Department found a definite sales resistance to the use of a front-mounted magneto which, rightly or wrongly, the average motorcyclist regarded as being too prone to unreliability in the rainy British climate. Whether this was a valid criticism is debatable, but it is certainly true that Ernie Nott probably lost the 1930 Ulster Grand Prix on this account and since the 1931 races were to have rear-mounted magnetos as a precaution, the production machines would have to follow suit. This was in line with John Pugh's dictum that the firm should be seen to be racing what were basically production-type motorcycles.

Instead, the 1931 production machines had separate front-mounted dynamos driven from the nearside engine mainshaft by a chain enclosed in a one-piece metal pressing that also covered the primary chain. The magneto was driven by an enclosed chain from the camshaft on the offside of the engine. The nearside chain cover had no oil bath and, as a new departure, the oil tank was separate and mounted on the offside of the saddle tube. However, the most noticeable design departure on all models, except for the DT machine, was a new oil pump of much greater capacity than hitherto, mounted horizontally at the bottom of the timing chest.

In January 1931, experiments were carried out with a new cylinder head with radial exhaust valves and parallel inlet valves. This design of head reduced the total number of moving parts in the rocker gear, provided increased cooling for the exhaust ports and resulted in a useful power increase, as a result, over the pent-type, parallel-valve cylinder head without the extra complication of the radial-valve head. The new valve arrangement was adopted for the 1931 works racing 500s.

As mentioned in the previous chapter, production of the JAP-engined 250 and 350cc models had been stopped and a new all Rudge-Whitworth designed and produced 249cc engine was on the stocks undergoing testing. This had a bore and stroke of 62.5 × 81.0 mm and a cylinder head which was virtually a scaled-down version of the 350 radial

For 1931, Rudge-Whitworth introduced this radial-valve head, with its six rockers, on its production 350s and new 250 production machines.

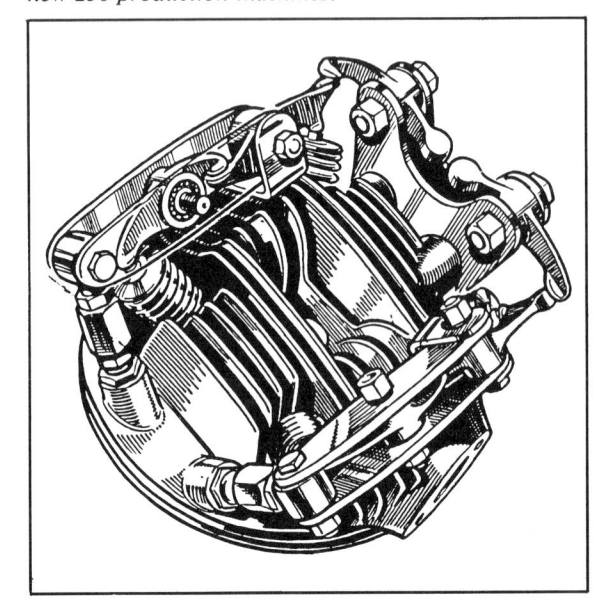

four-valve cylinder head. The downstairs department was quite different, however. In contrast to the firm's normal practice of racing new designs in the Isle of Man TT and then, if they were successful, putting them into production immediately afterwards, with the new 250, financial considerations made it imperative to put it into production right away. Like the other machines in the range it also had a rear-mounted magneto, which was driven by chain from the end of the camshaft. The overall weight of the machine was reduced from that of the 350 to make it competitive with rival lightweight Villiers-powered machines.

The first racing 250 single was assembled in March 1931. The same month news came through of a Rudge winning the 500cc Class of the Australian TT Races, while at home the competition season had started well for the company with Rudge riders winning the Birmingham Club's Victory Trial and the White Trophy in the Cotswold Cup Trial.

It had been a busy time for the Rudge-Whitworth development team, which as usual was attempting to do too much in too short a time. George Hack, Ernie Nott, Jim Dalton and 'Goldflake' Wills had sweated away their time, week after week, over the winter months in their attempts to get all of their latest developments ready in time for the coming season. Not only had they to cope with the entirely new 249cc four-valve unit, but, in addition, there was the revolutionary experimental 250 supercharged twin and the new semi-radial cylinder head

Joe Francis testing his Dirt Track Rudge out in the paddock at the Crystal Palace Speedway in April 1931.

for the 499cc models, to say nothing of trying to conjure up more power from the works racing 350.

On top of that there had been a major design change which could not be tested on the bench, the repositioning of the magneto behind the engine and its effects on handling. The result of this change was soon to be felt on the racing machines, where added to a mistake in the frame assembly it was to provide severe handling difficulties.

It was this far too extensive development programme that was to lead to the Rudge racing team's downfall in 1931 events, as much as the improvement in performance of the machines of its most serious racing rivals, the works racing Nortons. By the time April had arrived, though, and with it the North-West '200' Road Race to provide a preliminary trial for the new TT machines, the new 250 four-valve single Rudge racer was churning out some useful power. The new 60° V-twin experimental 250cc racing engine, on the other hand, was something of a disappointment. With four-radial valves per cylinder, centrally located 14-mm sparking plugs and a skew-gear-driven magneto mounted longitudinally on the timing cover, it was producing only some 80 per cent of the power of the single cylinder. The fact that a T-junction manifold had to be employed and the fact that it was at an early stage of development, no doubt largely accounted for this, but its performance was still well below par. Development had just not had time to produce useful results and there was no time to proceed with it any further with the racing season now about to start.

On the racing 350 single some experiments with 'bell mouths' on the carburettor had provided a useful increase in performance; Ernie took such a modified model out on the Leamington Road and unofficially attained a speedometer reading of over 100 mph (161 km/h). The machine also steered well, or so it seemed, which was pleasing. On the test bench a brake mean effective pressure (bmep) of 182 lb/in^2 (12.3 bar) and a peak power output of 28 bhp at some 6400 rpm were recorded. All this was achieved using straight exhaust pipes without megaphones.

Since the test bench and road results tallied, it seemed that on the 350 at least, four more usable horsepower could be relied upon compared with the previous year when the firm had achieved its 1-2-3 Junior TT victory.

The Rudge racing team set out for Ireland with models, however, which had never run on the road. Ernie Nott had the new semi-radial 500, while Graham Walker took a 350 with normal straight-through exhaust pipes; megaphone exhausts were considered a little too tricky on acceleration for the North-West '200' circuit and, anyway, George Hack wanted to save these as a surprise for 'The Island'. Tyrell Smith was down to ride the new 250 single. This, of course, was a completely unknown quantity so far as road performance was concerned. In fact the race itself proved totally misleading so far as predicting the Rudge team's form in the Isle of Man; but the riders and George Hack could not know that.

The three capacity classes in the North-West '200' Road Race were run together. George Hack's belief in the new 250 model was amply justified, for Tyrell Smith not only broke the lap record for the class on it from a standing start, but completely demolished it on the second circuit with a lap at 66.89 mph (107.71 km/h), only to drop out on his 12th lap, when leading comfortably, with a seized small end.

This failure was another consequence of an overloaded development department not having the time to tie up the various loose ends of a basically excellent design. For some astonishing reason the Rudge-Whitworth Drawing Office in specifying the same oil pump for the new 250 as was fitted on the larger models, somehow had managed to reverse its direction of rotation in the final specification. The race mechanics being unaware of this had coupled up the scavenge and feed pipes in the normal fashion. As a result Tyrell Smith had actually gone through the whole of the practice period for the North-West '200' and most of the race itself on the small quantity of oil above the level of the oil-tank return pipe, which in this instance was actually feeding and not scavenging!

In the 350cc event Graham Walker led from start to finish. Tim Hunt (Norton), his only real opposition, dropped out in the first lap with a broken rocker. Graham, therefore, had what was virtually a high-speed tour at a record average speed of 67.39 mph (108.52 km/h). His fastest lap, however at 69 mph (111.11 km/h) was nine seconds slower than Hunt's 1930 record, so his motor was not unduly stressed.

The 500cc race developed into a duel between Ernie Nott and Stanley Woods (Norton), with Harry Meege (Norton) entering the argument now and then in the early stages. At three-quarter distance, however, Stanley got a slow puncture and, although he continued gamely to finish second, he was no real opposition to Ernie who went on to win at a record average of 72.97 mph (117.50 km/h), his third successive victory in this race.

As a consequence of the opposition effectively disappearing from the scene in the two larger classes, the problems about to appear with the 350 and 500 racing machines failed to reveal themselves before the Isle of Man TT. The result of all this was that the Rudge works team went off to the Manx races with an air of confidence that was soon to be rudely shaken. The party comprising Nott, Tyrell Smith, Walker, George Hack and three mechanics, took no fewer than 26 machines of varying capacities with it to 'The Island', including an experimental bronze-headed 500cc racing model and four 1930-type Senior machines, purchased by an oil company and to be ridden by Leo Davenport, Ralph Merill, G.W. Wood and J.J. Byrne. Much to the embarrassment of the 'official' Rudge teamsters these models proved to be actually faster than the 1931 racing machines.

From the weather viewpoint, the 1931 TT practice period was one of the worst in memory. To have to prepare for three races in only 12 days would have been bad enough, but this was now reduced to only nine. The first three, Thursday, Friday and Saturday, consisted respectively of thick fog, cold with a terrible wind and one of the worst downpours the Island had ever seen. So far as the Junior Race was concerned, the only guides to form for the bookmakers were Friday's laps, in which Jimmy Simpson (Norton) made the running in 32 min, with Jimmy Guthrie (Norton) second in 32 min 7 sec and Graham Walker (Rudge) third in 32 min 10 sec. This said nothing about their true racing potential.

The Monday practice period was wet and misty, and Tuesday's consisted mainly of thick mist. The weather became patchy on the Wednesday and then on the Thursday a fine day arrived at last! With it the Rudge team got a nasty shock when Stanley Woods (Norton) broke the Junior lap record by 9 seconds in 31 min 12 sec, at an average speed of 72.5 mph (116.75 km/h), with Freddie Hicks (AJS) second in 31 min 23 sec and Jimmy Guthrie (Norton) third in 31 min 34 sec. They were all a lot faster than the Rudge camp would have wished.

The Rudge teamsters were getting worried about their prospects, for the occasional fine practice periods had led to the awful discovery that the new frame on the 350 and 500 models was wrong. Why this had not revealed itself in the North-West '200' remained a mystery. What they did know was that, whereas their steering was as steady as a rock on the straights, as soon as they cranked over for a corner, the machines persisted in trying to come upright again unless held down by sheer brute force!

The team frantically tried different fork links, fork-link lengths, engine plates for varying the power units' positions and everything else that could be thought of, but to no avail. It wasn't until the Chief of the Drawing Office arrived two days before the Junior Race scheduled for the Monday, that the mystery was solved. Glancing casually over Graham Walker's model, which was ready to hand, he suddenly exclaimed: 'Well, well, you've got the wrong chain stays!' They had. Someone had blundered.

With all their various troubles and the awful weather in practice, the Rudge team had been unable to gain reliable data regarding the most suitable gear ratios for the race, so working on the principle that the law of averages would give Monday fine weather, the 350 works Rudges were each given a 5.2:1 top gear instead of the usual 5.4:1. This was the last of a train of errors that spelt the team's downfall.

At 5:00 am on the Monday — race day morning — it was pelting down with rain. However, by 7:00 am it was sunny and at 10:00 am, when the first machine was sent on its way, it was crystal clear, despite a tremendous headwind blowing down the Mountain against the riders.

The result of all this showed in the first lap, which was completed with Jimmy Simpson (Norton) leading at 73.88 mph (118.97 km/h); Ernie Nott on the leading Rudge lying fourth at 72.73 mph (117.12 km/h) and Tyrell Smith right back in 12th spot. Graham Walker was even more disadvantaged due to his bulk and 15 stone (95.5 kg) weight, and even further back still. On Lap 3, Tyrell Smith dropped out on the Mountain with a dry petrol tank, which was subsequently traced to a defective float chamber.

Meanwhile, Simpson (Norton) still led, at 74.36 mph (119.47 km/h), with his team-mate, Stanley Woods, second at 73.51 mph (118.39 km/h), but Ernie Nott was now only third at 73.29 mph (118.02 km/h) much against the odds, a mere 27 seconds behind Woods. At the same time, Graham Walker had moved up into eighth spot on the other Junior works' Rudge.

On Lap 4, a major pile-up at Glen Helen caused Ernie Nott to drop back a place, but Graham Walker was now sixth. Nortons now held the first three places with Hunt in the lead. Lap 6 saw Ernie Nott take over third spot again from Stanley Woods and Graham Walker moved into fifth spot, and these were the positions that the two Rudge men finished in, Ernie Nott coming home third at 72.37 mph (116.54 km/h) and Graham Walker fifth at 70.98 mph (114.30 km/h). Definitely a case of mind winning over matter in view of the circumstances.

After the Rudge Junior Race débâcle, the telephone wires fairly hummed with harsh words between the Rudge team's Island headquarters and the Rudge-Whitworth works in Coventry. It had been too late to alter the 350 machines' rear ends in time for the Junior Race, but it would be really rough if the Senior Rudges couldn't be modified in time for Friday's race.

Fortunately there were no ultra-fast Nortons to worry about in the Lightweight Race line up and the new radial-valve 250 works Rudges had been functioning perfectly. There was some measure of justifiable confidence in them, therefore.

It had been Tyrell Smith's job to concentrate on the 250s during the practising periods and he had clocked up some really fast laps despite the terrible weather conditions. After a couple of tours on the Thursday at around the 60 mph (97 km/h) mark, he completed three laps on the Friday at an average of 65.6 mph (105.64 km/h), which was faster than the previous year's winning average. Then Ted Mellors (New Imperial) knocked Wal Handley's 1930 Lightweight lap record for a Burton, breaking it by 6 sec with a time of 33 min 7 sec, representing 68.60 mph (110.46 km/h) and on a foggy morning, too. This, however, did not disturb the Rudge teamsters unduly since 15-stone Graham Walker on the same morning, with all the disadvantages that his weight and bulk on such a small machine entailed, had gone round in 34 min 12 sec and 34 min dead, without trying too hard. Clearly, with Ernie Nott and Tyrell Smith in the saddle, the speed could be much higher. This was amply demonstrated to all and sundry on the second Thursday morning's practice, when much to everyone's astonishment (except the Rudge camp of course) Ernie Nott managed to put

Rudge-Whitworth had a disappointing 1931 Junior TT Race. Here is Ernie Nott after coming home third — the first Rudge rider to finish in this event.

in a couple of incredible record-breaking laps in 32 min 51 sec or at 69 mph (111.11 km/h); this during one of the brief, belated appearances of the sun. Meanwhile, Graham Walker managed to equal Ted Mellor's earlier performance of 33 min 7 sec from a standing start.

In later years when recalling the build-up to the 1931 Lightweight TT, Graham Walker said: 'Privately I was astonished by my lap times, I had scarcely ridden even a touring 250cc prior to race week.'

Graham was more than a little pleased with his lap times during practice, because Team Manager George Hack had, quite rightly, as Graham was the first to admit, strongly opposed his entry in the Lightweight Race. It was only due to John Pugh's sporting intervention that he appeared amongst the runners on race day. Logically, George Hack's attitude was correct, for what was the point in slaving away for months to pare off a bit of weight here and a bit of weight there and gain a few extra horsepower by diligent tuning and then go and throw away this advantage over the opposition by overloading the poor little machine with a rider weighing some 50 to 60 lb (23 to 27 kg) more than either of his team mates.

By the use of Elektron magnesium-alloy crankcases and gearbox shells, plus other weight saving dodges, the 249cc works' Rudges had had their weight reduced to around 255 lb (116 kg). With an 8.5:1 compression ratio and peaking at 6800 rpm, the output of the engines powering the machines of Graham Walker and Tyrell Smith was about 17 bhp, giving them a top speed of about 84 mph (135 km/h). Ernie Nott's machine, on the other hand, had a compression ratio of 10.6:1 and was fitted with various special bits and was some 4 mph (6 km/h) faster. All three machines had rev counters mounted on the sides of their fuel tanks, with the critical upper engine speed limit clearly marked in red to remind the riders of the dire threats made by George Hack against any of them who dared approach, let alone exceed, an engine speed of 7000 rpm.

The previous year, one of the problems that emerged during Junior practising on the 350s was the existence of a certain dangerous range of engine speed just below the consistent figure the riders sought to maintain throughout a race. If the engine speed was kept within that danger zone for any length of time, the valve springs broke. When this happened, the little H-section hardened piece between the transverse rockers flew out and that was that. The fitting of rev counters on the Junior as well as the Lightweight machines enabled the 'danger zone' to be avoided to some extent. They were not always reliable, though, and riders were obliged to depend a lot upon the 'feel' of their motors when this dangerous rev range was about to be approached.

Other features of the new 249cc radial-valve Rudges included 14-mm central sparking plugs — which were infinitely easier to remove than the 'candle-like' 18-mm jobs used formerly—and 8-in (203-mm) diameter front and 6-in (152-mm) diameter rear coupled brakes. Strangely, the front and rear wheels were very slightly out of track yet, according to the riders, they were the most perfectly steering machines they had ever ridden.

After all the machines had been wheeled into the ACU tent on the Tuesday morning, the day before the race, a torrential downpour ensued. This brought a profound depression upon the riders as they had looked forward to race-day weather that was at least dry. They needn't have worried though for the Wednesday morning dawned sunny, despite being cold and windy. What a relief!

The instructions from George Hack were that Ernie Nott was to win and Tyrell Smith was to back him up, and Graham Walker was to cover the laps in his own time. The first lap was sensational. No fewer than eight runners had beaten Wal Handley's 1930 lap record from a standing start. At the end of it Ernie Nott led, having carved nearly 2 minutes off the record and had become the first Lightweight rider to lap the TT circuit at over 70 mph (113 km/h), with a speed of 70.81 mph (114.03 km/h). Tyrell Smith lay second at 69.18 mph (111.40 km/h) and Graham Walker third at 68.58 mph (110.43 km/h).

According to the late Graham Walker, the machines of Ernie Nott and Tyrell Smith, despite their disparity in top speed, both employed a top gear ratio of 6.2:1. Graham's on the other hand was lower, at 6.32:1, to take account of his greater bulk and, as he later admitted, his tendency to let his engine speed get too close to the dreaded valve-spring-breaking danger period on occasions.

Getting back to the race, Lap 2 saw 'Nottie' set another sizzling lap record in 31 min 46 sec or 71.28 mph (114.78 km/h), while Tyrell and Graham occupied second and third berths at 69.30 and 69.05 mph (111.59 and 111.19 km/h) respectively.

The new 249cc Rudges handled so well that they could be cranked over at seemingly impossible angles for the day. In fact when Tyrell Smith came in to fill up on his third lap, he actually had a tuft of grass stuck in the end of his handlebars from the roadside bank at Hillberry! On this lap Ernie Nott again broke the lap record! This time in 31 min 34 sec or 71.73 mph (115.51 km/h) and his average speed now stood at 71.73 mph (114.83 km/h), over 2 mph (3 km/h) higher than that of Graham Walker who now lay second as a result of Tyrell's pit stop, Tyrell now being in third spot. The end of Lap 4 saw the order of the first three the same but no records broken.

All three Rudge teamsters came in to refill on Lap 5. Ernie Nott caused a bit of a stir when he arrived at his pit before Tyrell Smith had got underway again, having made up the 5½ minutes separating him from his team-mate at the start. From there on he was under orders to ease up a bit on the throttle and follow Tyrell round the circuit to conserve his engine.

Graham Walker was particularly relieved to come into the pits as he had a most unfortunate experience between Ramsey and Sulby. In an effort to reduce the wind resistance of his great bulk on so small a mount, he had unwittingly slid so far back that he went right off his rear pad and could not get back again, due to the fact that he had caught the fly of his leather breeches in the zip of the pad! As he later recounted: 'For some miles I was stretched out like a flying "V", frightened to stand on the exhaust pipes lest I broke their fragile clips.' He, in fact, only did so when the pain from the friction between his thighs and the rear lifting handle became unbearable and the approaching Parliament Square made rapid braking essential.

The end of Lap 6 saw the same order for the first three places. Meanwhile, the spectators had become bored with what seemed like a Rudge-Whitworth procession and many had drifted off to the local hostelry. Then, as so often happens, the last lap provided a sensation. Ernie Nott had stopped at Governor's Bridge. The Rudge pit team was in a state of furore! What had happened!

The explanation was not long in coming, when shortly afterwards, Ernie Nott crossed the finishing line in acute pain, with his right hand hanging down bleeding. Apparently, with a 4-minute lead and victory in sight, his inaccessible exhaust tappet had slacked off. With nothing but a big shifter and a magneto spanner to make an adjustment that would have taken seconds with the right tools, he had actually ridden the last part of the race *holding the pushrod in the tappet!*

His grit and courage paid off, for despite losing third place to Ted Mellor (New Imperial), by the gallingly small margin of 26 sec, by finishing he enabled Rudge-Whitworth to take the Manufacturers' Lightweight Team Prize. Graham Walker, who finished 2 minutes after Ernie, assumed the latter had won and only realised that he himself was the winner on time, when Tyrell Smith burst into the finishers' wash and brush-up tent where he was dousing himself, to congratulate him.

The official result was that Graham Walker had won at 68.98 mph (111.08 km/h), with Tyrell Smith second at 68.26 mph (109.92 km/h). Plucky Ernie Nott, in fourth place, averaged a fine 66.72 mph (107.44 km/h) and this third lap time of 31 min 24 sec stood as a new Lightweight course record.

Ernie Nott's 1931 Lightweight TT engine was raced with a Duralumin connecting rod and experimental big end. The rod was not bushed, but worked directly on the crankpin. The scheme proved a good one, so the 350 and other 250 racing engines were later similarly equipped. On test after the TT, Nott's 250 engine churned out some 19 to 20 bhp at 6700 rpm on the brake, while his 500 was producing 33 bhp at around 5900 rpm on a standard 7.25:1 compression ratio using a No 53 main jet.

On the Continent success was only intermittent. Ernie Nott won the 350cc Class of the FICM Grand Prix and Graham Walker the 250cc Race. In the German Grand Prix at the Nürburgring, the Rudge teamsters found themselves dicing with some very fast Austrian and German machines, including a supercharged water-cooled Puch two-stroke in the 250cc Class, and Ernie Nott had to be content with third place on his atmospherically-aspirated radial-valve 250 Rudge. Things went slightly better in the 350cc Class which Tyrell Smith won, but the marque had to be content with third place in the 500cc category.

The next port of call was the Dutch TT at Assen. In the 250cc Class a supercharged Puch led initially, followed closely by Ted Mellors (New Imperial) with Tyrell Smith in close pursuit on the Rudge. When the Puch fell out, Mellors lost his pace and Tyrell managed to slip past him to win at 72 mph (115.94 km/h). In the 350cc Race, Ernie Nott gained a second place, while Graham Walker finished third in the 500cc event.

Practice for the 1931 Belgian Grand Prix at Spa, went quite well until the weather turned wet and the downpour continued until the Sunday, race day. This, combined with the fact that a 24-hour car race the previous week had left slippery rubber all over the circuit, resulted in extremely difficult and dangerous riding conditions. As it was, Graham Walker was forced by the foul weather and the general inaccessibility of the circuit to hire a lorry to take the works Rudges from the garage near the hotel at which they were staying, the several miles uphill to the other hotel near the circuit where the machines were being weighed in and examined by race officials on the Saturday prior to the race.

All in all, the Belgian Grand Prix was another anti-climax. The 350cc Class was undoubtedly the race of the day with Jimmy Guthrie, Stuart Williams and van Hammersveld sailing round in close company, until after a fill-up Williams dropped out and with only three laps to go Hammersveld retired, so that Ernie Nott was able to finish second to Jimmy.

The next big event for the Rudge team was the Ulster Grand Prix in September and it was to be its last chance to make a real impact on the 1931 road racing scene. The 500cc Class of the Ulster proved to be an epic struggle between Ernie Nott and Stanley Woods on the Norton. On the first lap they were level pegging. Lap 2 saw Ernie just one second ahead. He had increased this to 2 seconds on Lap 3. Then on Lap 4 Stanley put in a sensational lap at 89.67 mph (144.40 km/h) to lead by 24 seconds. He increased this lead to 35 seconds by the end of Lap 5. But then Ernie turned on all the taps and knocked this lead down to only 6 seconds on Lap 6 (at half-distance). He could not sustain this effort though and Woods went on to win with Ernie second at 85.86 mph (138.26 km/h). Wal Handley finished fourth on another Rudge.

Nortons had things all their own way in the 350cc Race and the best a Rudge could do was third place at 78.65 mph (126.65 km/h) in the hands of Patrick Walls. Fortunately, nobody could catch the little radial Rudges in the 250cc event and they scored a fine 1-2-3 victory. Ernie Mitchell came home first at 72.63 mph (116.96 km/h) and made the fastest lap at 76.16 mph (122.6 km/h), with the two other Rudge-Whitworth teamsters, Paddy Johnston and Charlie Manders, following him home in second and third spots at 72.25 and 69.56 mph (116.34 and 112.01 km/h) respectively. Mitchell also won the 250cc Handicap and the newly instituted Governor's Trophy.

On the track at Brooklands, where high power was more important than supreme road holding, several Rudge machines performed well throughout 1931, largely in the hands of amateur riders. W.C. Marshall (Rudge) made an early appearance at the track, on Saturday, April 18 1931, and gained third place in a 350–1000cc Solo Handicap at the Second BMCRC Race Meeting of the year. Then at the Third BMCRC Race Meeting, postponed from May 16 to Saturday, June 27, L.R. Courtney (499cc Rudge) finished second in the 350–1000cc Solo 3-Lap Handicap and the Non-Trade Members' 3-Lap Handicap.

All this was a prelude to the first Brooklands' win of the year by a Rudge-Whitworth when, at the Fourth BMCRC Members' Race Meeting at the track on Saturday, July 18, L.R. Courtney came home first in the 350-1000cc Solo 3-Lap Handicap at a creditable 96.33 mph (155.12 km/h). In the process he completed one of his flying laps at over 100 mph (161 km/h) to win one of the coveted Brooklands' Gold Stars. At the same race meeting, Associate Members of the BMCRC were timed individually over flying laps of the track's Outer Circuit on their own machines and G.L. Reynold (499cc Rudge-Ulster) won the 500cc category at 80.85 mph (130.19 km/h).

At the Fifth BMCRC Brooklands' Race Meeting of 1931, on Saturday, August 15, all three of these Rudge riders achieved some success. W.C. Marshall finished second in the 350-600cc Solo One-Lap Scratch Race for the Prestwich Cup, with L.R. Courtney third. In the Over 245cc 3-Lap Handicap Race, Courtney finished third. Then, in the Non-Trade Members' 3-Lap Handicap for the Driscoll Cup that followed, W.C. Marshall won on his 499cc Rudge-Whitworth at 94.68 mph (152.46 km/h), with G.L. Reynold second.

The high performance of L.R. Courtney'a 499cc Rudge-Whitworth had not gone unnoticed by the Brooklands handicappers and when he entered the 90 mph 3-Lap Handicap for the Baragwanath Cup, he found himself pushed back to the 6 seconds start over scratch mark, in company with Lance Loweth and his potent 498cc Loweth-JAP. From this position he could only manage a third place.

The 1931 Brooklands racing season ended excellently for the Rudge marque, with a fine win by L.R. Courtney on his 499cc model in the Non-Trade Members' 3-Lap Handicap at 100.21 mph (161.37 km/h), with W.C. Marshall (499cc Rudge-Whitworth) second. In the course of the race Courtney completed one of his flying laps at the excellent speed of 105.52 mph (169.92 km/h). At the same meeting, W.H. Rigg (499cc Rudge-Whitworth) brought his machine home into third place in the 350-1000cc Solo 3-Lap Handicap.

Meanwhile, on the Continent other successes had included the winning of the Royal Grand Prix of Rome, the Italian TT Race and the 250cc Class of the Swedish Grand Prix.

On the trials front the firm had been extremely successful. In the Scottish Six Days' Trial, Jack Williams, Bob MacGregor and A.D. Stewart, riding 499cc solos, took the team prize and Rudge riders gained three Silver Cups and a Gold Medal. Later, in the International Six Days Trial, the six Rudge riders in the event all won Gold Medals for first-class performances. The official Rudge-Whitworth works team, comprising Betty Lermitte, Jack Williams and Bob MacGregor, carried off the 500cc Manufacturers' Team Prize with the loss of no marks. In addition, the two Rudge riders making up the Dutch national team, won the International Vase Trophy. The firm also won the Travers Trophy, which was subsequently put on display at the Rudge-Whitworth depot in Coventry, until it was stolen in a smash-and-grab raid!

The year 1931 was one of considerable experimentation by the Rudge development department. These experiments showed considerable variety. For example, John Pugh still harboured yearnings to attack the world motorcycle speed record despite his earlier problems with officialdom, in the form of the FICM and the ACU, with regard to his original proposal of a large capacity machine for the attempt — larger in capacity than the then recognised maximum capacity class of 1000cc. He had designed and built a supercharged 500cc flat-twin engine, with this in mind. The cylinder heads were standard 250cc Rudge four-valve types. The crankcase split vertically into front and rear sections with a large finned sump, and a magneto, chain-driven from the front engine mainshaft, was mounted above the unit. Unfortunately, the financial difficulties facing the firm at that time prevented the scheme going forward and so another one of John Pugh's imaginative projects was still-born.

In August 1931, experiments with the three-valve 500cc cylinder head were revived, but the large inlet valve still proved unsatisfactory, valve bounce setting in at a much lower speed than with the four-valve engines. It was therefore discontinued. Instead, the four-valve head was redesigned with smaller inlet valve guides and with the inlet port entry region machined to an oval shape to take account of the obstructions to gas flow created by the valve stems and guides. This resulted in an immediate power increase from 33 up to 36 bhp at 6000 rpm. This development was then standardised in the 1932 season's racing engines. Later in September, George Hack ran a careful series of experiments with different fuels in collaboration with ICI. The experimental 500cc engine with oval inlet port, when tested on RD1 racing fuel, which contained ethanol and methanol, together with acetone and up to 10 per cent water, showed a power output of 40 bhp at 5800 rpm using a compression ratio of 10:1. Running on straight methanol an output of 42.5 bhp was obtained on the same ratio, at 6000 rpm.

In its efforts to get out of its financial difficulties and keep on an even keel, Rudge-Whitworth started to supply all sizes of engines and gearboxes to other manufacturers, as proprietary parts under the trade name 'Python'.

Since the Grindlay-Peerless factory was only just a few hundred yards down the road from the Rudge-Whitworth works in Coventry, it was not surprising that it was one of the first British marques to use the Rudge-made Python units. Cotton and HRD soon followed, then AJW, Dunelt, Newmount (which based its machines on the German Zundapp but used tubular frames) and the Rex-Acme concern.

In Holland, Eysink and New Rap machines were being fitted with four-valve Python engines. Similarly, in Belgium, Bovy motorcycles (designed by Albert Bovy), Lady machines (designed by Lambert van Ouwerkek), and Escol motorcycles all used Rudge-made Python units. Meanwhile, in France, other machines using them included Durandal, Follis and Motopedale.

With its strong reputation over the years in Italy, it was also not surprising that, since they were made by Rudge-Whitworth, the new proprietary Python engines were to prove popular with Italian motorcycle manufacturers. These included Aquila of Turin, Doglioli & Civardi, Linx, Meldi and Miller-Balsamo, which used Python four-valve engines built under licence by Miller. The Swiss firm, Universal, also employed them, while in Germany the leading users were Imperia and Zundapp. In fact, Imperia actually called its 346cc Python-powered machine 'The Ulster', thereby betraying the origins of its power unit. Back in England, Chris Tattersall started using Python engines in his CTS racing machines.

At the Motorcycle Show in November 1931, Rudge-Whitworth showed its 1932 range of machines, which had quite a number of important design changes. Thus, an aluminium-alloy, oil-bath chaincase was used in place of a primary cover, which entailed a longer engine driveshaft and gearbox mainshaft. This enabled the 350 and 500cc clutches to be redesigned to cope with the increased power outputs. A central stand operating lever was also introduced for the first time that enabled a Rudge machine to be lifted on to its stand using only a finger.

Fully radial cylinder heads were now fitted on the 499cc Ulster and Replica models, despite the fact that the previous year's works racers had radial exhaust valves and parallel inlet valves. Unlike the Ulster models, the 350 and 500 Replicas went back to using front-mounted magnetos. The road-going Special now had a deeply valanced front mudguard and was available with the option of a chrome and black petrol tank instead of the customary black

Ernie Nott leading the 1931 Lightweight TT before having his tappet locknut slacken off in the last lap and losing him the race. He managed to finish fourth, holding it in place by hand!

finish or maroon for the tank and mudguards. An optional extra for trials' enthusiasts was an unswept exhaust pipe with high-level silencer. A new introduction was a 250 Replica model similar to the 1931 works racing machines. This had its magneto mounted behind the engine. Also, the road-going 250 could now be had with either coil ignition, magneto or Maglita.

The firm continued its line of sidecars from the previous year, but reintroduced its 8-cwt (407-kg) fabric-covered trailer, with commercial users in mind. Meanwhile, John Pugh decided to relinquish his function as Chief Designer to Frank Anstey, later to become famous for his work on the Ariel link-type spring frame. Pugh still kept a watching brief on developments, however.

With 1931 such a sales disaster, Rudge-Whitworth would have to make considerable economies if it was to stay in business. One obvious area was racing. All the leading motorcycle factories involved in racing were finding it increasingly expensive. So it was that only ten works racing machines were developed by Rudge-Whitworth for the 1932 season: three 250 radials, each producing around 19 bhp at 6700 rpm on a 9:1 compression ratio; three 350 radials developing 25 bhp at around 6300 rpm on an 8:1 ratio; and four 500 semi-radials which gave 39 bhp on a 10:1 compression ratio at around 5900 rpm. All these figures were achieved on petrol-benzol. Running on methanol, the 500 gave a further 2.5 bhp. Why the works' racing 500s still retained the older semi-radial exhaust-valve arrangement in contrast to the full-radial arrangement of valves on the road-going 500 Ulster and 500 Replica at this time is not clear. However, the reversion to a front-mounted magneto was intended to improve machine handling compared with the previous year's machines and this change also involved going back to using the vertical type oil pump. This was given increased capacity to cope with the additional oil feed provided to the cam gear. After Graham Walker's experience of having his mudguard breaking free when his forks bottomed during the 1931 Ulster

Left to right: Ernie Nott, Wal Handley and Tyrell Smith, with one of the new 250 radial-valve Rudge racers each was to ride in the 1932 Lightweight TT Race on the 'Island'. Note the fishtail silencer added to satisfy the letter of the law during road tests.

Grand Prix, rubber buffers were now attached to the lugs on the front forks.

Reverting briefly to the new oiling arrangements, or rather, the old oiling arrangements, the 1932 500cc racing machines also had the sinusoidal grooves in their oil pumps reversed so that compared with the 1930 racing version the pump ran in the opposite direction. Because of the confusion this caused amongst race mechanics in the Rudge team familiar with the older 1930 arrangement, in one race they inadvertently connected it up incorrectly, causing Tyrell Smith's retirement with a seized small end due to oil starvation.

The 1932 competition year started well for Rudge-Whitworth with a win in the Australian Grand Prix. At home on the trials' front the new 250 made its presence felt in no uncertain manner with wins in the Kickham Memorial Trophy and the 250cc solo classes of the Bemrose, Wye Valley and Cambrian events. In the Scottish Six Days' Trial the Rudge team won the Manufacturers' Team Prize, taking the premier solo award and three Silver Cups in the process. Then, in the International Six Days' event, the British 'A' Team, which was made up entirely of Rudge riders and comprised Graham Walker, Jack Williams and Bob MacGregor, carried off the Manufacturers' Team Prize, and two other Rudge-Whitworth riders in the six-day trial won Gold Medals.

The astonishing thing was that at this time motorcycle sport was flourishing, despite the precarious state of the British motorcycle industry. It seemed rather like fiddling while Rome burned!

In the run up to the Isle of Man TT Races, the Rudge racing team gained wins in both the 350 and 500cc classes of the North-West '200' Road Race and a second place in the 500cc category. Traditionally a testing ground for the TT events, it helped answer a few questions raised by recent experiments with carburation and so on. Quite a few experiments had been carried out with the use of air filters, for example, but these were dropped in favour of bell mouths since, although filters increased engine-cylinder bore life, they quite definitely impaired power outputs.

Jack Williams — the 'Cheltenham flier' — on the 250 Rudge which he rode in the 1932 Lightweight TT race in the Isle of Man.

For the Island races, a full Rudge-Whitworth works racing team was entered, comprising Graham Walker, Tyrell Smith, Ernie Nott and Wal Handley, who had signed up to ride in all three races. Their machines were clearly faster than their principal opposition in the Junior and Senior events, the works' Nortons of Woods, Guthrie and Simpson, but they still suffered from unsolved handling difficulties with the exception of the 250s which steered like a 'dream'. The change in engine and frame design for the 1932 races had merely raised the speed at which the tendency for the machines not to crank over into a bend occurred. Apart from these changes, the 500cc machines were also now fitted with megaphone-type exhausts for the first time, which initiated a trend in motorcycle exhaust system design, although I would not think that Rudge-Whitworth would claim to have been the first to use such systems. Certainly some car racers had been experimenting along these lines in the past.

In practice, Jimmy Simpson made the fastest lap in both the Senior and Junior categories, with Wal Handley second in each case. Whereas, however, Wal's best Senior lap was only 2 seconds slower than Jim's, he was a full 45 seconds slower in the Junior. It looked as if Nortons would have it all their own way in the Junior Race, but that it would be a close tussle to the finish in the Senior. As it was, things happened exactly the other way round.

In the Lightweight TT, bad luck again hit Ernie Nott. On his sixth lap he had overtaken Leo Davenport on his rapid New Imperial and led by some 22 seconds when, at Ramsey on his last lap, he blew his engine up and was forced to retire. Wal Handley was afflicted with sparking plug trouble throughout his ride in this event and was forced to change plugs three times in the course of the race. Despite this he held the lead for more than half the distance and finished third at an average speed of 69.86 mph (112.50 km/h) and set up a new lap record at 74.08 mph (119.29 km/h). Graham Walker, after a slow first lap, due largely to his leg problems and his bulk, rode his usual reliable race to finish second at 70.07 mph (112.83 km/h).

Stanley Woods (Norton) won both the Senior and Junior TT Races. There was no way that the Rudges could hold the excellently steering Nortons in these events, especially over a course like the Mountain Circuit where the requirement for good handling was paramount. A slight edge on power was no compensation for brilliant handling in the opposition. Thus it was that in the Senior Race Nortons achieved a 1-2-3 victory with Ernie Nott fourth, Graham Walker sixth and Tyrell Smith eighth.

Wal Handley had been running third for the first three laps, when he had a very nasty and potentially lethal crash, near Kirkmichael. Falling and damaging his spine, he lay in the road for some seconds, unable to move out of the way of any of the approaching machines. Furthermore, his Rudge lay alongside him with its engine revving its heart out and petrol from its tank flooding all over the place, so that Wal was soaked in it. Paralysed as he was, he was terrified that the pair of them would go up in flames before help could reach him. Fortunately, this did not happen! Wal's accident, though, which

Graham Walker on board his 1932 Lightweight TT 'works' Rudge, on which he finished second, having won the race the previous year. The schoolboy in the picture is his son, a very young Murray Walker.

landed him in hospital for a spell, put him right out of commission so far as road racing was concerned for the foreseeable future.

The Junior Race, which had been run earlier, produced a better result for the Rudge team, Wal Handley running Stanley Woods a close second all the way home and finishing second at 76.36 mph (122.96 km/h). Tyrell Smith came home third at 74.02 mph (119.19 km/h) and Graham Walker fifth.

Even on the Continent, where circuits tended to be faster and should have favoured the greater speed potential of the Rudge machines, the marque failed to obtain a single win in a major Grand Prix during 1932. However, in local European races some useful successes did result and the company was able to record wins in the 500cc Class of the Austrian TT and the 250 and 500cc Classes of the Spanish TT. But it was hardly a road-racing year for the Rudge-Whitworth team to rejoice in! Nearer to home came some useful wins in the 250 and 500cc classes of the Leinster event over in Ireland.

Immediately after the Isle of Man Races, the Rudge team decamped to Assen for the Dutch TT, but the best it could do was a third place in the 350 and in the 500cc races. In the Belgian Grand Prix at Spa, the team gained slightly better results, Charlie Manders gaining second place in the 250cc category and a further second place being gained in the 500cc Class.

After all the gloom and despondency on the sales and racing fronts, the Ulster Grand Prix provided a more hopeful end to the 1932 season. In the 350cc Class, Alec Mitchell (Velocette) took the lead at the end of Lap 1 and stayed in front for four laps, with Tyrell Smith (Rudge) back in fifth place. Then a whole host of retirements amongst the leaders put Tyrell into a lead with Charlie Manders on another Rudge second. They finished the race in those positions at 77.89 and 76.75 mph (125.43 and 123.59 km/h) respectively. Arthur Simcock, the Australian rider, held up the Rudge prestige in the 500cc Class by finishing third at 82.00 mph (132.05 km/h) and, in the 250cc Race, H. Pilling finished third on his Rudge-engined CTS.

The parlous financial state of the company was again reflected in a miserable production figure of 2500 machines for 1932. This led to immediate economies including the unavoidable cancellation of a number of projects that might have helped restore the Rudge-Whitworth concern's trading position in the longer term if they had been persevered with. Amongst these was a new Speedway machine to replace the existing model, which was now not really competitive with its JAP-engined rivals.

A new development that was carried on with was a 175cc ohv four-valve engine having many parts in common with the 250 power unit, including the crankcase and flywheels. It was designed to take advantage of the lower road tax on under-200cc machines in some European countries. Marketed under the name Python, it was sold to a number of French, German and Italian motorcycle manufacturers.

Rudge machines had more encouraging race results at Brooklands during 1932, largely in the hands of private owners. In fact, the very first motorcycle race of the Brooklands racing year was

Prince George of Kent chatting with Wal Handley prior to the start of the 1932 Senior TT Race.

Wal Handley astride his 1932 Senior TT Rudge.

Ernie Nott leaping Ballig Bridge in the Isle of Man, during the course of the 1932 Senior TT Race in which he finished fourth.

won by such a rider, V. Chirney. He took his machine home to a win in the first heat of the Clubmen's 2-Lap Solo Handicap, the first event at the First BMCRC Brooklands Race Meeting of 1932 on Saturday, March 19. His speed was 82.20 mph (132.37 km/h). Being the winner of the faster of the two heats he automatically won the Final.

In fact the whole of that meeting was devoted to Clubmen's Races and other successful Rudge riders that day included: Chirney again in the Clubmen's Flying Lap event, which he won with a speed of 86.62 mph (139.48 km/h); W.H. Rigg on a 249cc model, who won the BMCRC Non-Trade 3-Lap Handicap at a fine 80.20 mph (129.15 km/h) and W.C. Marshall on a 499cc machine who came third in the same event; J.A. MacDonald; and A.T.H. Debenham, the last two both on 499cc models.

At the Second 1932 BMCRC Brooklands Meeting, on Saturday, May 7, H.J. Bacon (499cc Rudge-Whitworth) had a fine win at 98.43 mph (158.50 km/h) after starting 10 seconds from scratch in the All-Comers' 3-Lap Handicap. He gained his Brooklands Gold Star for his first 100 mph (161 km/h) lap of the Outer Circuit in a Race, in the process.

At the same race meeting, Fergus Anderson, who was later to make a name for himself as a rider of Moto-Guzzi works racing machines, rode Ernie Nott's old 200-miles-in-2-hours record-breaking 499cc Rudge-Whitworth home to a comfortable win at 97.85 mph (157.57 km/h) in the 'Gold Star' 3-Lap Handicap.

As part of its policy to encourage the crowds to return to Brooklands to watch motorcycle racing, the BMCRC, at its Third Brooklands Race Meeting of the year, staged two so-called Brooklands Grand Prix over a special circuit designed to simulate road-racing conditions. H.J. Bacon took his 499cc Rudge-Whitworth to victory in the 10-lap Senior event at an average speed of 67.03 mph (107.94 km/h), taking the main prize of the day.

Other notable Brooklands performances by Rudge-Whitworth riders and machines during 1932 included a win at 81.24 mph (130.82 km/h) by Jack Lafone and his 346cc model at the Fourth BMCRC Meeting on June 25, and wins in the Junior and Senior Brooklands 100-Mile Grand Prix Races by Tyrell Smith (346cc Rudge-Whitworth) and Ernie Nott (499cc Rudge-Whitworth) respectively, at speeds of 70.64 and 72.54 mph (113.75 and 116.81 km/h), at the Fifth BMCRC Race Meeting of the year on Saturday, July 23.

When the Rudge-Whitworth models for 1933 were introduced at the end of the year, they showed little obvious change from their 1932 predecessors. Coil ignition had been discontinued on the 250, and the 500cc Special was now advertised as the 'Silver Vase' model in memory of its notable ISDT success. The 350 and 500cc Replica and Ulster models now followed John Pugh's dictum that production machines should be basically similar to the firm's racing machines, by once again having front-mounted magnetos and semi-radial exhaust valves; and the 350cc Replica machine retained its fully radial valve disposition.

The Dirt Track machine had now been dropped completely, opening up the flood gates to JAP-powered machines on British cinder tracks. Two sidecar models had been dropped and two new ones launched. In fact one of the new ones was actually called the 'Launch' and followed Rudge-Whitworth's traditional practice of styling its Sports sidecars like small boats — hence the name.

This, therefore, was the range with which Rudge-Whitworth hoped to see itself out of the trough of a very nasty economic depression.

13

The Syndicate (1933-1935)

Nineteen-thirty-three marked a crucial stage in the history of the Rudge-Whitworth concern. Things were very bad financially and drastic action was needed to force a recovery in the firm's fortunes. On January 15 1933, the company issued an official statement of policy to the effect that it would not be competing officially in any events during the coming year and that instead would redirect the efforts of its former racing department into development work for production machines.

This bombshell did not come as a surprise to the members of the racing team, who had already been informed of its contents before the official announcement was made, and Graham Walker, Tyrell Smith and Ernie Nott had already got together to discuss the future. They agreed to a suggestion from Graham that they should form a syndicate, to be known as the Graham Walker Syndicate, to which Rudge-Whitworth Ltd had agreed to lend its racing machines as and when required by the Syndicate, throughout the 1933 racing season and to provide all necessary support within its power in the form of mechanics' time and spare parts, providing it did not involve the company in any additional expenditure or detract from its principal aim of manufacturing production motorcycles for sale to the public. Graham, Tyrell and Ernie would then enter themselves on these machines as private entrants under their own names. It was racing on the cheap, but it could be done successfully and anyway there was no other option open to them in the circumstances, being Rudge-Whitworth employees, if they wished to race in international events. They would of course be allowed time off to do this, as had also been agreed from the outset.

Unfortunately, a number of less honest Rudge-Whitworth employees, seeing the vast quantity of special racing bits lying around the unattended Rudge racing department during the winter months, also took it into *their* heads to 'borrow' from the firm, albeit on a more permanent basis. The technique involved attaching these parts to wooden packing cases and floating them down the River Sherborne, which flowed through the works site. Then, walking empty-handed through the security controlled front gate where they were liable to be searched, the parts could then be collected later, unobserved, from downstream. Quite a large number of unusual Rudge racing parts came into private possession in this way.

It is interesting to note at this stage that it was not only the Rudge factory that suffered from large-scale pilfering by its own employees at this time, for the bankruptcy of A.J. Stevens, manufacturer of the AJS motorcycle, was largely due to internal stealing, albeit on a much larger scale. Rather a sad thing to have to recall, but that was one of the insidious effects of the depression years — the crime rate went up.

Getting back to the firm's financial state, John Pugh, obviously upset by the turn of events, shortly after the 'no-racing' statement had been issued, announced his resignation as Managing Director of Rudge-Whitworth Ltd, in favour of Frank G. Woollard. He continued, however, to take an interest in the firm still in the role of Chairman of the Board.

Things got steadily worse financially during the next two months and, on March 27 1933, H.H. Graham was called in to act as Receiver and Manager of the company. The way the company's profits had plummeted over the preceding years is shown by the fact that a surplus of £11,690 at the end of 1929 had, by the end of 1933, been turned into a deficit of £33,503. It was not a happy situation to be in and there were no easy options available in seeking a way out of it in view of the state of the market at that time.

Production, however, did not cease with Mr Graham in charge of things, but the range of machines was drastically reduced and became considerably more rationalised than hitherto, to help cut production costs in a diminishing sales situation. All of the 350 models were discontinued, as was also the 250 Replica. There were, however, detailed improvements to the rest of the machines.

The 500s were now fitted with Ransome & Marle big-end assemblies in place of the former Rudge-made components. On the standard model, the Special, the R & M assembly used ¼ in × ⁵⁄₁₆ in (6.35 mm × 7.94 mm) rollers in place of the less substantial, needle rollers used on the Rudge component, while the Ulster R & M big end employed three rows of ¼ in × ¼ in (6.35 mm × 6.35 mm) rollers instead.

The main changes on the Special were to its

appearance. Thus, on the offside of its petrol tank, the knee grip now covered the pivot of its hand-operated gear change, while on the same side of the machine a metal shield covered the gearbox. The oil tank on both models had been reshaped from a fairly well-rounded design to a squarish configuration, to match that of the battery, mounted on the offside of the saddle pillar, which was now enclosed in a special cover. The whole effect was to produce greater lateral symmetry to the machine.

The greater standardisation of component design between the two 500s was clearly directed towards reducing production costs. It was carried further, too, since both models now had almost identical bottom halves to their engines, forward-mounted dynamos and rear-mounted magnetos. The only obvious difference, apart from the more substantial big-end assembly and heavier-duty main bearings on the Ulster, was the use of the prefix 'S' on the Special's crankcase and the prefix 'U' on that of the Ulster.

Upswept exhaust pipes and high-level Burgess-style tubular silencers were now standardised on the Ulster and on the 250 Sports model, which also had a slipper-type piston, a sports cam and stronger valve springs.

By May 1933 the Syndicate had available for its use not only the 1932 250 and 350 racing machines, but also four new 500cc racing machines, which had been based on carefully assembled 500 Replicas, each having a separate oil feed to the camshaft and two of whose engines were fitted with aluminium-bronze cylinder heads.

An attempt was also made to try and solve the handling problems that had beset the racing 500s over the previous two years. This involved extending the petrol tank rearwards to fit round the peak of the saddle, but without actually increasing its capacity and thereby bringing the centre of gravity further back. The same logic was involved in lightening the front forks and hub. The effect of these changes was to produce an overall machine weight for the 500s of around 300 lb (136.4 kg), despite the fiting of a heavier and more substantial clutch.

In May, the Syndicate entered the new 500s in the North-West '200' over in Ireland and had its only major win of the year, although it was of course not to know it at the time. In the Senior TT in the Isle of Man though, because there had been no development work in the Rudge-Whitworth racing department during the previous seven months on account of the firm's financial state, the new 500s proved to be totally outclassed by the works' Nortons. The best the syndicate riders achieved was a fifth place for Ernie Nott, a sixth for Graham Walker and a ninth place for Tyrell Smith. In the Junior TT there was only one Rudge rider who finished, Fernando Arunda, a private entrant from Spain, who came home 12th.

In view of the fact that Rudge-Whitworth Ltd no longer produced the 250 Rudge-Replica, it would not allow the Syndicate to enter the Lightweight TT Race, although this view changed the following year. However, a private owner, the Irish rider, Charlie Manders, had entered his own very fast 1932 250 Rudge-Replica in the event and had succeeded in finishing third at the very creditable speed of 69.17 mph (111.38 km/h). He was, in fact, only 7 min 45 sec behind Sid Gleave, the winner at the finish, whose ride to victory marked the debut of the very remarkable 'Mechanical Marvel' Excelsior. Other finishers on 1932 Rudge 250 Replicas included Les Martin, L.P. Hill and Bill Kitchen, who came home in eighth, 13th and 14th positions respectively.

The Graham Walker Syndicate also managed, in the 500cc Class, to gain a third place in the Dutch TT at Assen and a second in the Belgian Grand Prix at Spa. In the 500cc Class of the Ulster Grand Prix, the works' Nortons of Stanley Wood and Walter Rusk dominated the race, but, biding his time carefully, on the last lap Ernie Nott took his privately entered 499cc Rudge into third place at a respectable 82.69 mph (133.16 km/h).

In the 250cc Race there were four Rudges entered, including that of Charlie Manders who succeeded in finishing second behind Charlie Dobson on the works New Imperial. Manders averaged a fine 76.22 mph (122.94 km/h) and amazingly, considering the company he was in, set up the fastest lap of the race at 78.26 mph (126.02 km/h), thereby showing that Rudges were still a force to be reckoned with in the Lightweight category. Colin Taylor's CTS, powered by a Rudge-made 249cc Python engine, took third place at 74.45 mph (119.89 km/h), another fine performance against works opposition. As it turned out, Manders might well have won the race but for the fact that his front forks had been giving him trouble and had forced him to slow down somewhat. Even so, Dobson won by the narrow margin of only 39 seconds. It was a close-run thing!

An indirect Rudge victory came in the Italian Grand Prix at Rome, the last international road race of the year. In the first part of the 500cc race, three Moto Guzzis of relatively new design led the field, but were experiencing handling difficulties. They led in close formation until one of them ran off the track and a little later another stopped at the pits to change plugs. This let Carlo Fumagalli and his Python-engined Miller-Balsamo into second place. Then Sandri, on the leading Moto Guzzi found his machine's bad handling increasingly difficult to cope with and Fumagalli slowly forged ahead never to be repassed.

At Brooklands in 1933 there was consolation for the Rudge marque with a number of excellent performances by Rudge-Whitworth riders and their machines, private owners in the main.

The track's racing calendar started with *The Motor Cycle* magazine's Clubman's Day Race Meeting on Saturday, March 25. In the course of this, Rudge machines shone in a number of events, the most outstanding achievements being G.H. Kynaston's 91.68 mph (147.63 km/h) on his 499cc Ulster in the Clubman's Flying Kilometre Trials and D.W. Ronan's winning of the Clubman's 'Mountain' 3-Lap Passenger Handicap at 50.83 mph (81.85 km/h), on a similar model with sidecar attached.

Mountain Circuit Races were now becoming a regular part of the Brooklands' 'scene' as the track authorities (including the BMCRC) tried to attract back the crowds of former years to the circuit. So it was that at the Second BMCRC Monthly Race Meeting of 1933 on Wednesday, April 26, D.W. Ronan took his 499cc Rudge, this time in solo trim, to its second 'Mountain' victory in a 5-Lap All-Comers' Handicap. In the Flying Kilometre Trials at the Fourth BMCRC Meeting of the year, riding the same model, he clocked 98.11 mph (157.99 km/h), to win the 500cc Class of that event. At the same race meeting, he also secured second places in two further 5-Lap All-Comers' 'Mountain' Handicap Races.

During the course of *The Motor Cycle* magazine's 'Cup Day' on July 15 1933, Ronan finished second on his Rudge in the Wakefield Cup 10-Lap Mountain Handicap, while B.A. Chevell (also on a 499cc solo Rudge) came third in the Driscoll Cup Five-Lap Mountain Handicap.

Other successful riders of Rudges at Brooklands during the 1933 racing season included Jack Lafone and Jock West. The latter, who rode Ariel machines on grass circuits for the author's father during the 1930s, made his name in road racing on BMWs and later on AJSs when head of sales at Associated Motorcycles during the 1950s. It was in 1933 that he gained his Brooklands Gold Star by covering a 100-mph (161-km/h) lap of the Brooklands Outer Circuit in the course of a BMCRC event riding a Rudge and also took fifth place in the Senior 100-Mile Brooklands Grand Prix riding the same machine, on Saturday, July 29, that year.

The very much reduced range of four machines, two 250s and two 500s, was that with which the company started the year 1934 under the guidance of a Receiver. Sir Edward De Stein had become the new Chairman of Rudge-Whitworth Ltd in place of John Pugh, whose health was growing steadily worse, no doubt because of the trauma that was afflicting his company.

In addition to the modifications and improvements already described, the 1934 models each had a new design of rear mudguard. The previous hinge-up rear type had been dispensed with in favour of a two-part mudguard, the shorter front section behind the gearbox being fixed and the rear longer portion quickly removable after undoing two nuts. Another change was that electric lighting equipment was now a standard fixture, rather than an optional extra, as on all modern machines. This, however, did put up machine prices when most other manufacturers were succeeding in keeping theirs down. Thus the 500 Special now cost £7 more than in 1933 at £63 10s, the 500 Ulster at £73 10s was £8 more expensive, while the 250 Standard model which cost £52 10s was £6 dearer. The new 250 Sports model was priced at £55 10s.

In February 1934, despite its 'no-racing' stance and under strong persuasion from Graham Walker, the company produced a new cylinder head for the racing 500s using existing patterns. This comprised an aluminium-bronze cylinder head skull, with shrunk-on cast aluminium alloy cooling fins, with excellent heat dissipation characteristics. Although this was much lighter than the existing cylinder heads, the differing thermal expansion rates of the two head materials produced problems that could not be resolved with the resources to hand. The project was therefore dropped.

A month later a more successful bi-metal development occurred with the design of a front brake drum with shrunk-on, aluminium-alloy finning. With this a thin-finned, aluminium-alloy ring was cast up and shrunk to a standard brake drum. This improved front brake cooling, but added extra unsprung weight to the front suspension. This was compensated for to some extent by the adoption of a lighter, high-tensile steel, front-wheel rim.

With private owners having done so well in 1933 on 1932 Rudge-Replicas in international road racing, the firm produced, for the use of the Graham Walker Syndicate, a batch of three new racing 250s very similar to the 1932 Replicas but incorporating the new Ransome & Marle big end assemblies. Two new 500s were also produced, similar to the machines supplied to the Syndicate in

1933, but again with the difference of having the new R & M big ends. The new bi-metallic, finned, front-brake drum was also fitted to both sizes of racing machine for 1934. The new 500s also had Duralumin pushrods fitted, whereas the 250 racers retained the steel variety.

Another feature of the larger machines was the fitting of carburettors with some 10° of downdraft. Between each carburettor and its cylinder head was interposed a 1-in (25-mm) long extension piece. This provided a smooth transition from the circular bore of the carburettor to the oval entrance to the inlet port in the cylinder head itself. In an effort to improve the poor steering of the 500s, the front-fork links had been changed, while an extra rubber buffer had been fitted to prevent the front-fork spring's telescopic cover from fouling the upper fork links.

Isle of Man time came round again and as in 1933 the Graham Walker Syndicate riders rode their Rudges as private entrants in the Lightweight and Senior Races. Since there were no 350 Rudge racing models available from the works in Coventry, and since Rudge-Whitworth no longer produced 350s, Tyrell Smith felt free to ride a works AJS in the Junior while Ernie Nott rode a Swedish Husqvarna twin in the same event.

With only two 499cc Rudge machines available to the Syndicate, it was decided that these should be ridden by Graham Walker and Les Martin, Ernie Nott riding a works Husqvarna instead. Without any works backing, however, from the development point of view, the Syndicate's task was hopeless and the best it could achieve was a sixth place for Graham. Les Martin's machine failed to finish. However, an interesting entry in this race was that of A.P. Hammersveld on a Rudge-engined (more strictly Python-engined) Eysink machine built in Holland.

In the Lightweight TT things were different. Graham Walker had managed to silver-tongue that well-known rider of the day, Jimmy Simpson, into joining the syndicate. Thus, Walker, Les Martin, Ernie Nott and Simpson would ride the Syndicate's 250 Rudges in the event. This combination provided a 1-2-3 Rudge victory as it turned out.

Jimmy Simpson, who had a reputation as an engine killer, due to his tendency to ride flat out as often as possible, in this, his 13th Tourist Trophy Race, after three second places, four thirds and seven record laps, added another fastest lap and a *first* place to his great riding record. Never before had he actually won a TT. The race itself was by no means a Rudge walkover. It was run off in bad weather conditions, with thick mist on Snaefell Mountain, and was marred by the fatal crash of Syd Crabtree, a well-known and popular rider.

Charlie Dodson (New Imperial) led on the first lap, with Ted Mellors (Excelsior) in hot pursuit and Jimmy Simpson third. Mellors slowed in the next lap and Jim moved into second spot with Ernie Nott, on his Rudge, third. Then both of them passed Charlie and stayed first and second until the end. Meanwhile Graham Walker brought his Rudge home into third place to make it a 'hat trick'. Starting in such awful racing weather, Jimmy Simpson had already decided to retire from the race as soon as practicable, his feeling for personal safety overcoming his normal hard-riding tendencies. Thus it was, that on the first lap, when his machine partially seized, he used this as an excuse to come into his pit with the object of

Cartoonist Jock Leyden's impression of Jimmy Simpson, the 'new' Rudge teamster for the 1934 Lightweight TT Race.

Jimmy Simpson, who scored a brilliant and popular Victory in the 1934 Lightweight TT Race on his 'Syndicate' Rudge.

retiring. His pit manager, however, much to his credit, refused to allow him to do so.

The final official result was a win for Jimmy Simpson at 70.81 mph (114.03 km/h) and a fastest lap at 73.64 mph (118.58 km/h), with Ernie Nott second at 69.76 mph (112.33 km/h) and Graham Walker, the 1931 winner, third at 67.67 mph (108.97 km/h), which gave Rudge-Whitworth the Manufacturers' Team Prize. Of the other two privately entered Rudges, owned by Manliff Barrington and Les Martin, Martin's failed to finish and Barrington's was a non-starter. In all, only eight of the original 24 starters finished and no records were broken because of the heavy rain. Regarding the mist on the Mountain, as Graham Walker remarked to Simpson after the race, 'It's all right for you little fellows, you can see under it!'

At the end of the 1934 racing season, having achieved his one remaining TT ambition, Jimmy Simpson retired from active motorcycle racing altogether, but still visited the 'Island' during TT week each year in the role of representative for a well-known oil company. After this great TT success, Tyrell Smith took himself off to the Continent, where he won the 250cc Classes of both the Belgian and German Grand Prix Races.

In the Ulster Grand Prix at the end of the season, in which the 250, 350 and 500cc Classes were run simultaneously, there was a massive 'pile up' involving some 14 machines and their riders and it was lucky no-one was killed. Fortunately, Ernie Nott managed to have the Irish 'piskies' luck on his side in this race and weaving his way through this mêlée went on to a fine win on his Rudge in the 250cc event at an average speed of 77.98 mph (125.57 km/h), making the fastest lap in the process at 78.84 mph (126.96 km/h). Private owner W.H. Hey took his 346cc Rudge into third place in the 350cc Class at 79.19 mph (127.52 km/h), while Tyrell Smith took his 'Syndicate' 500 into third spot in the 500cc category at 85.26 mph (137.29 km/h).

In addition to these successes, foreign Rudge riders had won the 500cc Class of the Targa Florio in Italy, and the 350 and 500cc Classes of the Estonian TT race, and had filled the first four places in the 500cc Class of the Greek TT Races. Meanwhile, Rudge riders had gained two Gold Medals in the 1934 International Six Days' Trial, one of them by Dr Galloway, a member of the successful British Silver Vase Team that year.

At Brooklands Track in 1934 Rudge-Whitworth machines continued to put up good performances in the hands of private owners, notably D.W. Ronan, Fergus Anderson, J.A. MacDonald and P.A. Refoy. At the BMCRC's Brooklands' 'Cup Day' Race Meeting on Saturday, May 12 1934, Anderson won the Three-Lap Handicap for the Staniland Trophy, at an average speed of 91.72 mph (147.70 km/h), while a month later at the BMCRC's June Race meet, Ronan won the All-Comers' Handicap over three laps, at the particularly fine average speed of 98.43 mph (158.50 km/h). Another useful 'over-90 mph' performance that year was that of Refoy in a Three-Lap Handicap at the BMCRC's July Race Meeting when he won at 93.62 mph (150.76 km/h).

Despite all these excellent performances, though, it was clear that the Syndicate was fighting a losing battle. Anyway, Graham Walker was finding his leg injuries too much of a handicap and decided to retire from active road racing at the end of the 1934 road-racing season. This effectively wound up the Graham Walker Syndicate, although by arrangement with the Rudge-Whitworth concern, Tyrell Smith was allowed to carry on racing one of the 250s during the 1935 season. John Pugh's worsening health had forced his retirement from the company on doctor's advice. Meanwhile, by the end of 1934, only some 2000 machines had been sold, as in the previous year.

There is no doubt that a lot of the success achieved by the Syndicate can be attributed not just to the dedication of Nott, Tyrell and Walker, but also to a lot of surreptitious help from George Hack and his team. It now devolved largely on his overworked, under-financed, design and development team to come up with a new range of Rudge-Whitworth machines for 1935, still with the Receiver, H.H. Graham, overseeing things and the threat of liquidation looming large. Under these conditions one could have expected the existing 1934 range to have been continued unmodified, but design improvements were carried out by George Hack with the urgent objective of improving sales. As early as July 1934 he had experimented with a water-cooled cylinder head for the Ulster, but this did not reach the production stage. The eventual 1935 Ulster model, although superficially resembling its predecessor of 1934, had at least two design differences. The cylinder head now had much more substantial rockers while its petrol tank, like that on the 1934 racers, now extended back either side of the nose of the saddle. Most of the redesign work that did take place, though, occurred on the 250s. These now comprised the 250 Sports model, introduced for the first time the previous year, and a completely new machine, the 250 Tourist model, which sported a two-valve ohv cylinder head; the very first production Rudge-Whitworth motorcycle to do so. The Sports model retained fully radial valves, but its oil tank had been moved from the front of the crankcase to a more conventional mounting on the offside of the saddle tube. Like its larger brethren, this model also had more substantial rocker gear. The dimensions of the engines on both 250s were the same at 62.5×81 mm, which gave a swept volume of 249cc.

The reason for going over from four to two valves on the Tourist model was that although the valve gear on each was exposed to the elements, the greater number of working parts involved with the four-valve arrangement entailed a far greater amount of maintenance. During the development of the new 250 Tourist, in November 1934, a Lucas Magdyno was tried out, but was eventually rejected and both 250 models for 1935 were fitted with Maglitas instead.

No new 250 competition machines were produced for the 1935 racing season, but ten racing 500s were produced for private owner use. Five of these were exported to Germany, while the remainder were sold to British riders. It is reported that one of these (Engine No 804) developed 42 to 43 bhp at 6000 rpm. One engine on test, which had been quietly but extensively breathed upon by George Hack, put out a healthy 48 bhp until it decided to blow up.

Now, although the steering of the production touring machines was satisfactory, at racing speeds it was far from right. Because of this, in March 1935, experiments commenced with the object of improving high-speed handling. One of the early developments was pneumatic damping for the front forks. This failed to go into production though. A second approach involved the development of a new, more rigid, front fork employing round rather than D-section blades and an integral forged bridge instead of the bolted-up variety used till then. The latter proved a considerable improvement and was adopted on the 1936 range of machines.

Later in the year (1935) the design department turned its attention towards improving the gearbox. The needle rollers around the outside of the fourth speed pinion were shortened to make space for a pig-skin washer designed to reduce gearbox oil loss.

For the 1935 Isle of Man TT Races, the factory supplied Ernie Nott and Jack Williams with the other two 1934 racing machines. This enabled them, together with Tyrell Smith, to enter themselves privately in the Lightweight Race and form what in effect was a Rudge team in the way that the Graham Walker Syndicate had done in the past. The race, which was run in poor weather conditions, provided Stanley Woods (250 Moto Guzzi) with a runaway win. The main interest for the crowds though was in the monumental scrap for second place, between Tyrell Smith and Omobono Tenni on the other Moto Guzzi; but in the fifth lap Tenni retired after a crash at Creg-ny-baa and Tyrell rode home into second place at 70.67 mph (113.80 km/h), 2 min 48 sec behind the winner, with Ernie Nott third at 69.37 mph (111.71 km/h) and Jack Williams fifth. L.P. Hill, another 250 Rudge rider, finished in ninth place. This result enabled

Tyrell Smith (Rudge) after finishing second in the 1935 Isle of Man Lightweight TT Race.

Rudge-Whitworth Ltd to take the Manufacturers' Team Prize for the second year in succession.

After the TT, Tyrell Smith did the rounds of the Continental Circus on his 249cc model. In the process, in the 250cc Class, he finished second in the Swiss Grand Prix and third in the German and Belgian Grand Prix Races.

An interesting machine entered in the Isle of Man Senior TT Race in 1935 was a Rudge-based rotary-valve-engined machine ridden by Alf Brewin, which unfortunately retired on its second lap. T.D. Cross, who was responsible for its design and development, had obtained from Rudge-Whitworth Ltd a set of engines (250, 350 and 500) of the Replica type back in 1933 and had replaced the cylinder and head of each with one of his own design fitted with a special rotary valve in which the axis of rotation lay parallel with the crankshaft. The valve was made gas tight by chamfering the phosphor-bronze sleeve ports and setting them over slightly, so that when the valve was in position the spring edges sealed the joint. Lubrication of the sleeve was effected by supplying a liberal quantity of oil to the surface of the valve and removing the surplus by means of a scraper device. The oil supply was controlled by the throttle, so that it increased with the degree of throttle opening.

To overcome the possible binding effects caused by cylinder pressure, a system known as 'controlled valve loading' was used. It employed gas pressure to provide the necessary effort to exert a force on the two halves of the valve housing and controlled the amount of this force by the principle of mechanical leverage.

Experiments were continued by T.D. Cross throughout 1936 with all three sizes of engine. No further efforts were made to produce a racing engine though, but some very useful results were obtained. A 247cc unsupercharged engine fitted with a nitrogen-hardened steel Cross-type valve and housing, and an aluminium-alloy cylinder, eventually developed a bmep of 165 lb/in^2 (11.2 bar) at 5350 rpm and 176 lb/in^2 (12.0 bar) at 6000 rpm, running on 65 octane petrol and a compression ratio of, astoundingly, 10:1! The high thermal efficiency of this engine was reflected in its very low fuel consumption of 0.35 lb/bhp/h (0.21 kg/W/h) at 4000 rpm.

A Rudge-Ulster fitted with a 349cc Cross rotary-valve engine achieved 25 bhp at 6000 rpm running on 65 octane petrol and also using a compression ratio of 10:1. This machine gave a fuel consumption figure of 99 mile/gal (35.4 km/litre), with the machine cruising at between 60 and 65 mph (97 to 105 km/h).

At Brooklands throughout 1935, it was the amateurs who upheld the Rudge marque's prestige. At the BMCRC's April meeting at the track, *The Motor Cycle* Clubman's Day, in the Clubman's Flying Kilometre Sprints, Javier de Ortueta (499 Rudge) clocked 93.21 mph (150.10 km/h) in the 500cc Class, while in the 350cc category W.J. Jenness (349 Rudge) recorded 81.05 mph (130.52 km/h). Then, at the Club's July Brooklands' Race Meeting, Ron Harris (499 Rudge), the scratch man, made a great effort to snatch second place in the first of the 'Round the Mountain Races', a Five-Lap Handicap. He was more successful in the second such event over 10 laps and won it at 69.51 mph (111.93 km/h).

A series of Grand Prix Races was held at Brooklands at the end of August under the sponsorship of *The Motor Cycle* magazine. Two major events fell to the Rudge marque, with C. Bayly (499 Rudge) winning the 50-Mile Clubman's Senior Grand Prix for the Dunlop Trophy at 75.04 mph (120 km/h) and Ernie Nott (249 Rudge), the 100-Mile Lightweight Grand Prix at 73.68 mph (118.65 km/h).

The Brooklands' year ended for Rudge-Whitworth with Tyrell Smith and Ernie Nott making a collaborative attempt on 250cc long-distance solo records on one of the new 249cc two-valve models. However, after lapping consistently for about five hours at around the 89 mph (143 km/h) mark, the engine blew up. Then, Ernie Nott brought out a 249cc four-valve machine which had been set up on methanol and attempted a 100 mph (161 km/h) lap of the track on it. Several laps were completed at around 99 mph (159 km/h) and then the connecting rod broke and, in the absence of spare parts, that was that.

By the end of 1935 *again* only 2000 or so machines

Left *A weary Ernie Nott (Rudge) after finishing third in the 1935 Lightweight TT Race in the Isle of Man.*

Right *This Rudge-based rotary-valve engined machine, developed by T.D. Cross, was ridden in the 1935 Senior TT Race.*

Right *The new competition 500 Rudge 'Ulster' on display at the 1935 Olympia Motorcycle Show.*

had been sold, largely due to the high prices being charged for them compared with rival makes, and the company was forced into liquidation. One of the creditors, however, the Gramophone Company, part of the EMI Group of Hayes in Middlesex, saw some future prospects in the firm and bought the assets from the liquidator. Thus, under quite new management, Rudge-Whitworth announced its new models for the coming year.

Apart from the new front fork, an all-black petrol tank with gold lining and a new transfer, plus fishtail silencers, the 500 Special was unchanged. The improved front fork was also adopted on the Ulster model which apart from having downswept exhaust pipes now, was also unchanged. Both 500 models could be supplied with upswept pipes if so desired. The new forks were also fitted on the four-valve Sports 250 while the 250 two-valve machine was renamed the Rapid, with little or no change to its specification apart from a few more fins on the cylinder barrel. An interesting change on the 500s was the fitting of the so-called Rudge regulator. This was fitted in place of the eight-day clock and was essentially a speedometer fitted with a rotatable indicator enabling the engine revolutions in any gear to be determined. Particularly significant was the fact that the prices of the Special and Ulster models were unchanged.

14

The end of the road (1936-1943)

The year 1936 started on a tragic note with the death of the former MD of Rudge-Whitworth, John Pugh. It was he, if anyone, who had been largely responsible for the company's success over the years. Some said he had died of a broken heart, for his now ailing business had been his life's work since the death of his son in World War 1. The firm under its new management gave scant attention to producing racing machines during 1936. One or two 250cc and 500cc racing machines were made but they did not differ markedly from the production sports models.

Several private owners continued to race either these machines or earlier Replicas in road racing. Amongst them were Harold Hartley, who rode Rudges in the Isle of Man TT Races every year (except 1953) from 1936 to 1954 with the obvious exception of the war years, and Harold Newman. But Tyrell Smith, the last of the original 'faithful trio' that made up the Syndicate, now raced a works' 250 Excelsior, as did also Charlie Manders.

At Brooklands Track in 1936, J.H. Greenwood (499 Rudge-Whitworth) gained a Gold Star with a lap at 104.63 mph (168.49 km/h). This was in the course of the Five-Lap All-Comers' Handicap for the FitzGerald Cup, at the BMCRC's Cup Day Race Meeting in May, which he won at a fine average speed of 100.1 mph (161.05 km/h). Other successful Rudge exponents at Brooklands that season, included L. Tooth, L. Good, R.A. Mair, W. Cox, Ron Harris, who finished second on a 249cc model in the 100-Mile Lightweight Grand Prix Race in July, and Dennis Minett. The latter came third in the Hutchinson Hundred that year at the very respectable speed of 97.85 mph (157.57 km/h) for the 100 miles of the race.

On the trials' front though, the factory was reasonably successful and did not rely entirely on amateur riders to uphold its prestige. Thus in the International Six Days' Trial, held that year in Germany's Black Forest area, three Rudge riders, Bob McGregor, J.A. MacLeslie and J.E. Edward, all Scotsmen, made up the winning British Silver Vase B Team. This was only the second occasion on which the ACU had selected a one-make team to represent Britain in the International Silver Vase Competition. On each occasion it had been Rudge-Whitworth and on each occasion the result had been a win.

With the Nazis rise to power in Germany and the subsequent building up of that country's armed forces, rearmament was a major talking point in official circles in 1936. As a result there were Government contracts open to tender to supply the British armed forces with motorcycles for despatch riders. By June 1 that year, Rudge-Whitworth Ltd had come up with 249cc and 499cc WD machines based on its ISDT models and incorporating the quickly detachable hubs used in that event. Unfortunately these had to be modified to meet WD specifications, so a rear stand was also now fitted. The models were submitted to the War Department for testing, but the use of uncaged rollers in the gearbox was rejected. As a result two new experimental models using Burman gearboxes were made. Meanwhile, the firm had presented the London Division of the Territorial Army Signals Regiment's new display team with three of the new machines, which were used at some 14 different displays and gymkhanas during 1936.

Having developed the WD models, the design department under George Hack now attended to

The 1936 500 'Ulster' engine with the rocker box cut away to reveal the rocker arrangement.

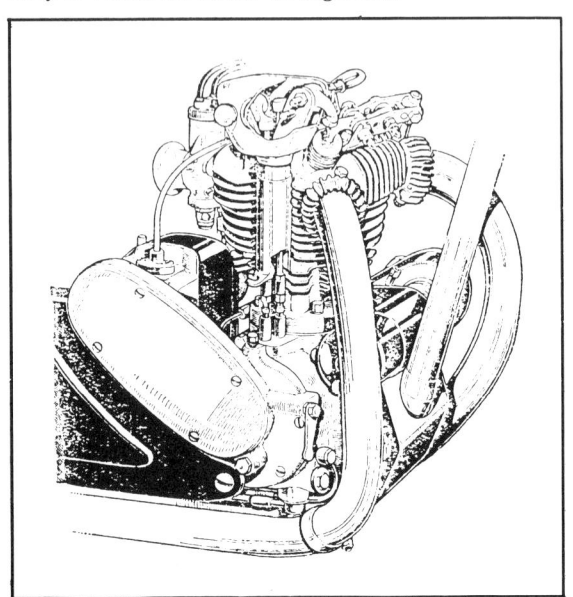

Chris Tattersall on his Python-(Rudge-)-engined 250 CTS, which he rode in the 1937 Lightweight TT Race. Chris was one of the most successful of the rider/constructors in the immediately pre-World War 2 period.

the road-going machines. First of all the needle rollers inside the fourth-speed pinion of the gearbox were now replaced with two Oilite washers to help cut down oil loss. The overdue decision was also made to enclose totally the rocker gear on the 500 Special and 500 Ulster models inside the aluminium-alloy covers and provide it with a positive oil feed. This was pressure fed through the rocker spindle, over the valve stems and down the pushrod tunnel onto the cam followers. This had the dual benefit of reducing rocker gear wear by isolating it from spurious abrasive road dirt and, at the same time, reducing wear on the cam and its follower. All this entailed giving the oil pump a larger capacity to cope with the increased flow rates required. Another design weakness was also dealt with at this stage, the use of solid pushrods. These had a nasty habit of wearing through the case hardening of the first rocker and then boring their way right through it. The answer was to use tubular steel pushrods with cup and ball contacts at each end. It was also anticipated that with the use of enclosed valve gear, the circulation of oil over the cylinder head would reduce the running temperature of the engine, so the oil feed to the rear of the cylinder was discarded. This was, in fact, to prove a mistake.

Changes to the cycle parts included the design, in July 1936, of a new rear wheel which, like its predecessors, was quickly detachable but ran now on Timken taper-roller bearings and had an integral brake drum and sprocket on the nearside. In so doing, the company aligned itself with the customary British motorcycle practice of having the brake pedal on the nearside and the gearchange on the offside of the machine. This in turn, now removed the unnecessary complication of a gearchange crossover shaft. As a result, the chaincase now had only one hole in its centre, for the foot-rest hanger. Furthermore, the rear wheel could now be removed simply by undoing the spindle nut, pulling out the spindle, knocking out the rear-wheel spacer and undoing the three bolts holding the wheel to the brake

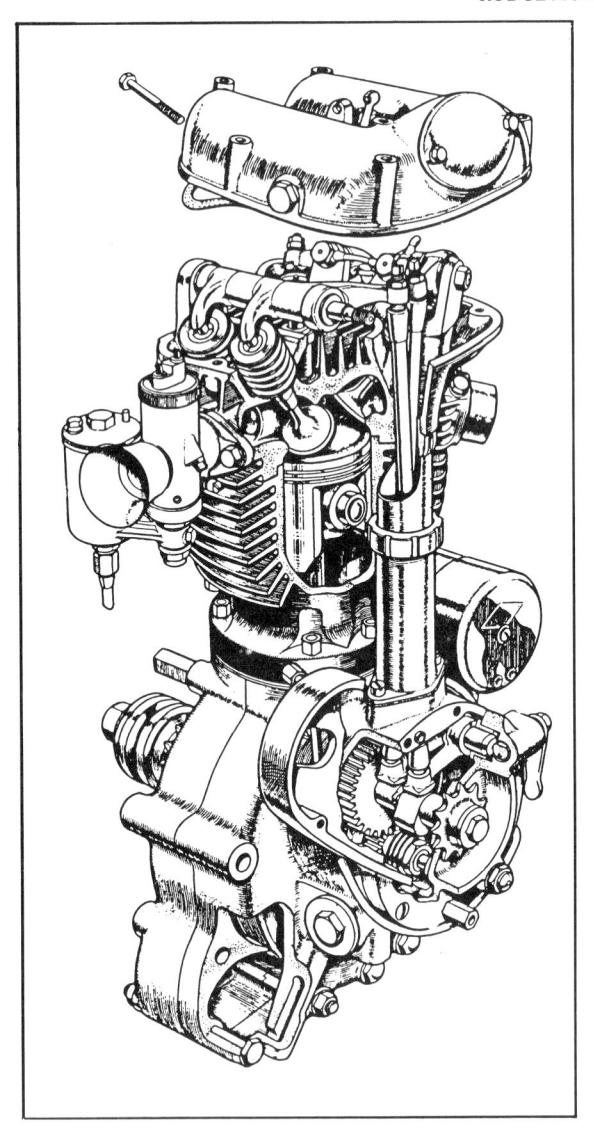

A part-cut-away view of a 1938 'Ulster' engine.

drum. It was now all very easy to do.

With only 2000 machines sold in 1936, for the third year in succession, it was decided to scrap the 250 four-valve model for 1937, retain the two-valve 250 and introduce a new 500 called the Sports Special. The new 499cc Ulster and Special models for 1937, with their enclosed valve gear, introduced to the public for the first time at the Olympia Motorcycle Show, now went into production, as did also the Sports Special. The latter was basically an ordinary Special but with upswept exhaust pipes, Ulster-type handlebars and front mudguard, and a rear mudguard with a straight valance fitting the space between the top of the upswept silencer and the mudguard. It also had two tool-boxes. A quickly detachable ISDT-type rear wheel was an optional extra on all three 500s, the three wheel-to-brake drum securing bolts now having been replaced by pegs to facilitate rear wheel removal.

Fortunately the new machines proved a hit with the buying public and by the end of 1937 some 3000 had been sold. Nevertheless the Rudge-Whitworth works at Coventry was working well under capacity so it was decided to sell it and remove motorcycle manufacture to Hayes in Middlesex, where EMI had its headquarters. Meanwhile Graham Walker had resigned from the Company and had taken up the position of Chief Editor with Temple Press Ltd's journal *Motor Cycling* on which he was to become one of the country's most widely known and respected authorities on motorcycling matters. Before he left Rudge-Whitworth he managed to persuade the firm to build three or four racing machines, one of which was supplied to Norman Croft for racing at Brooklands. The others were exported.

On the trials' front, an all Rudge-mounted team again represented Britain in the International Silver Vase Competition of the International Six Days' Trial in 1937, but this time unsuccessfully.

On the road-racing side, none of the original Syndicate teamsters now raced Rudge-Whitworth machines and the firm's reputation in the racing sphere was left in the hands of the faithful Harold Hartley and Chris Tattersall on his 249cc Python-engined CTS machines. In fact several 250 racing machines were built to special order in 1937 but that was the last occasion on which this happened. In the Lightweight TT that year, Harold Hartley retired, but he gained a 22nd placing in the Junior and finished 13th in the Senior, all on outdated works Rudges, which no longer had any serious development work carried out on them.

At Brooklands, things were slightly different. There, where performance counted and there were no restrictions on the type of fuel used, expensive alloy motors had little if any advantages and Rudge machines were going well in the hands of private owners. One of the most notable performances here took place at the last BMCRC Race Meeting of the year, when Dennis Minett, lapping at a steady 108 to 109 mph (174 to 176 km/h) won a 10-Lap Outer Circuit Handicap at the excellent average speed of 105.29 mph (169.55 km/h).

The new 1938 range, shown on the Rudge-Whitworth stand at the Earls Court Motorcycle Show in September 1937, was little changed from the previous year, except for price increases all round in the order of £4 to £5. Speedometers were now standard fitting, as the law now required them.

A new model was the 250 Sports machine, a tuned version of the two-valve Rapid, having a slipper piston giving a compression ratio of 7.7:1 in place of the 6.6:1 ratio of the full-skirted piston used on the touring model.

The discarding of the oil feed to the rear of the cylinder on the 500s had led to a number of owners writing in to the company during 1937 complaining of engine seizures. Because of this, for 1938 it was reintroduced. Another modification on all three 500s was the introduction of a slightly larger, rear bottom shackle bolt to take account of the strengthening of the bottom of the shackle.

Early in 1938 Rudge-Whitworth Ltd moved from Coventry to a new factory at EMI's Hayes site. Interestingly enough, although no machines were produced during the move, the year ended up with a total of 3000 being produced, which was not at all bad bearing in mind the circumstances. Before the move the company had decided to produce what nowadays would be classed as a moped, in the form of the Rudge autocycle and, as the firm produced no suitable engine, a 98cc Villiers Junior de Luxe power unit was used. This machine was to be the last powered Rudge two-wheeler to remain in production.

George Hack stayed in charge of motorcycle affairs at the Hayes factory, despite still living at Coventry and commuting each day from there. The result was rather a short working day for him, a situation which the new management was unlikely to tolerate indefinitely and he eventually left the company in 1939 to move into the world of aero engines. Before he did so though, he solved some of the problems associated with employing a heat-dissipating cylinder head on the Ulster model in a material lighter than aluminium-bronze. As will be recalled this work had started ten years before with the development of a bronze skull with shrunk-on aluminium-alloy finning. George eventually developed an RR50 aluminium-alloy head in which the problem of securing the head bolts was solved by using threaded bronze inserts.

Harold Hartley, seen here on the 250 Rudge he rode in the 1939 Lightweight TT Race, continued to ride this machine with a fair degree of success in most of the post-World War 2 Lightweight TT Races up to 1952.

Roland Pike who, with his brother, was one of the most successful 250 Rudge riders immediately before and after World War 2, is seen here aboard his 1939 Lightweight machine in the 'Island'.

The valve seats were of austenitic iron and were pressed into the head after being cooled in solid carbon dioxide and the head being heated in hot trichloroetheylene. This created an interference fit which at room temperature was about 0.005 in (125μm). Just for safety, though, the edge of each valve seating was slightly chamfered and the head material rolled over the chamfer. This was the head arrangement used on the 1939 Ulster models.

George also specified RR56 aluminium-alloy tubular pushrods. This was where his knowledge of metallurgy showed, for as the engine reached running temperature this material to all intents and purposes compensated for the thermal expansion of the cylinder head, so that hot and cold running clearances were essentially the same.

Despite the Ulster engine being designed primarily for road use, it was possible to order tuned engines at this time for £5 extra cost. These had special racing camshafts, a larger than standard bore carburettor, polished ports and a compression ratio in the range 7.25:1 to 7.5:1 depending upon the state of tune of the engine.

The Ulsters produced in the period 1937 to 1939 inclusive, were turned out with a ratio slightly under 7.0:1 using only one 0.080-in (2-mm) thick compression plate. With this removed and a 0.040-in (1-mm) plate substituted, the ratio was increased to about 7.5:1 and to 8.2:1 if taken out altogether.

The last Service Manager of Rudge-Whitworth Ltd, R.P. Ransom, although not exactly advocating the use of megaphones for road racing, gave some pointers as to the dimensions to be used with the qualification that they would provide a little more top speed at the price of some acceleration. For the 500s he suggested that megaphones, if used, should be 13 in (330 mm) long with a larger (outlet) diameter of 3¾ in (95 mm) and be fitted on to a pair of 1½ in (38 mm) diameter exhaust pipes that terminated 4 in (102 mm) forward of the rear spindle.

A number of private owners were finding that the 250 two-valve Sports machine was able to be quite

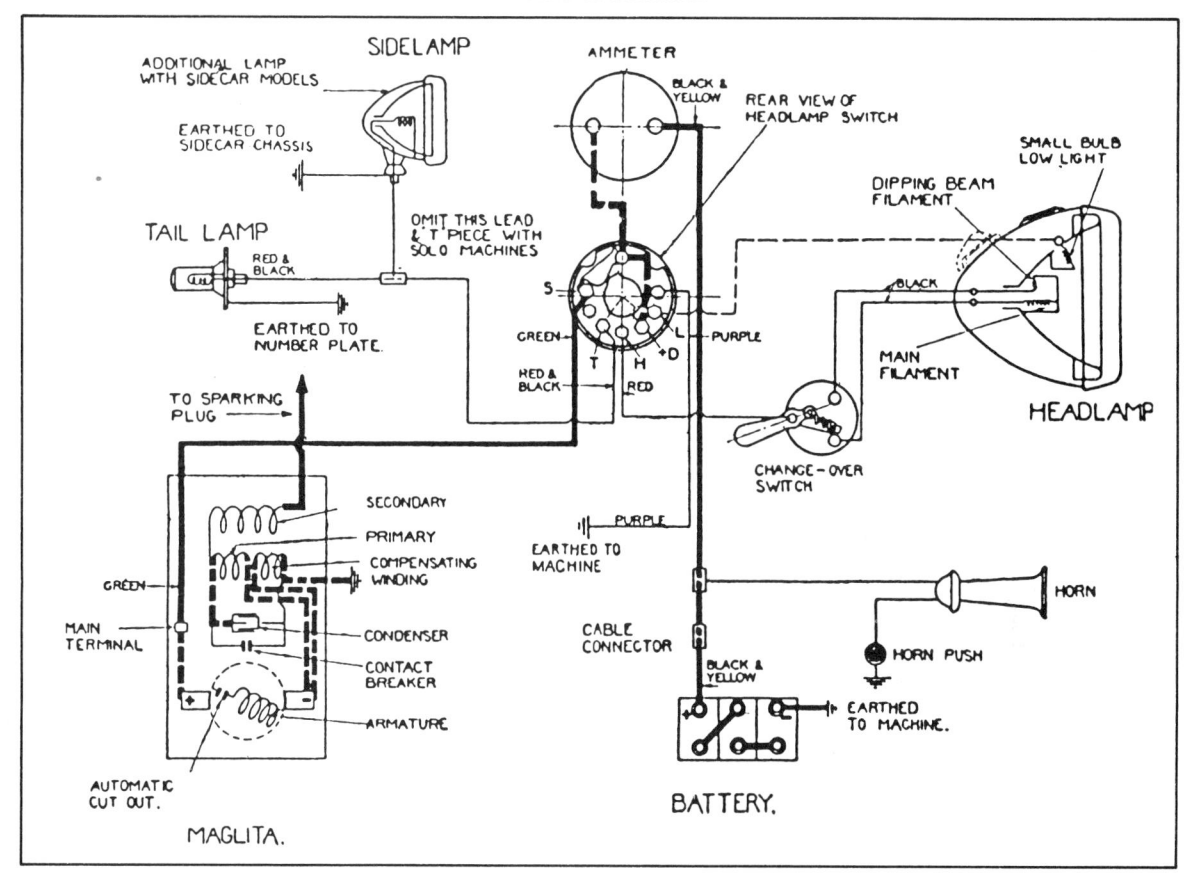

The Lucas Maglita combined lighting and ignition equipment wiring circuit employed on Rudge-Whitworth motorcycles.

The magneto chaincase assembly for the 1939 Rudge 'Ulster'.

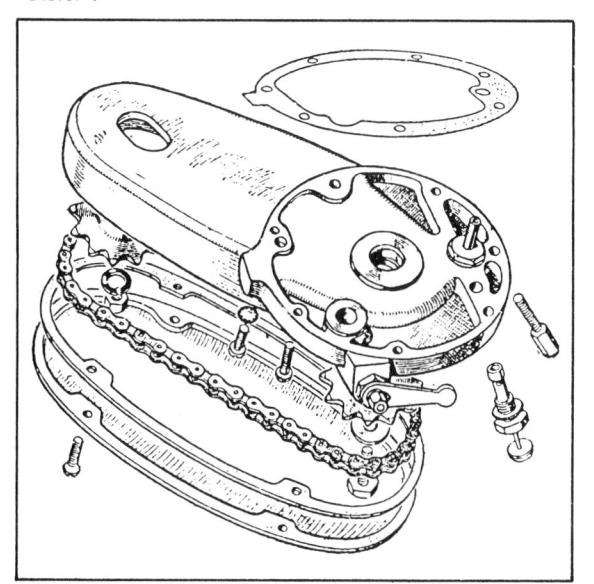

An 'exploded' view of the 1939 Rudge 'Ulster' crankshaft assembly.

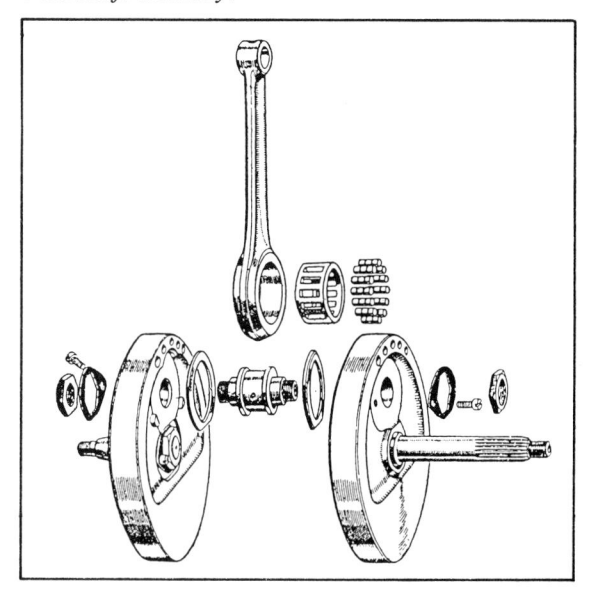

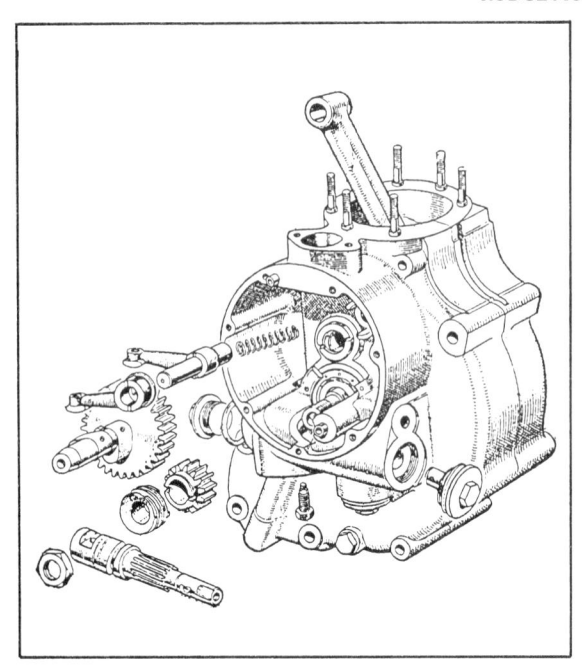

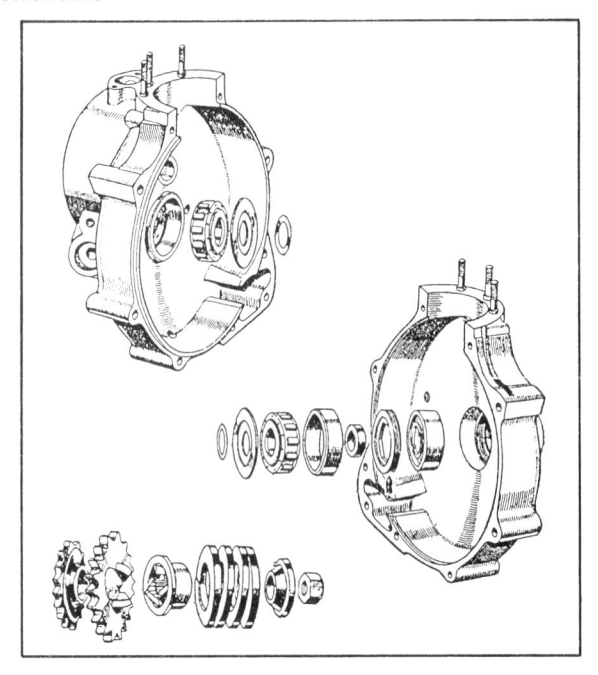

Above left *The timing side components of the 1939 Rudge 'Ulster' engine.*

Above *An 'exploded' view of the later type of Rudge 'Ulster' crankcase showing the order of assembly of the mainshaft bearings and other components.*

Left *Godfrey's, as this advertisement indicates, undertook the supply of Rudge spares when the marque ceased production. The firm only ceased this facility when it closed its Great Portland Street depot in the 1960s.*

competitive in road racing. Two of them were the Pike brothers, Stan and Roland, who first rode Rudges in the Island in the 1937 Manx Grand Prix.

In the 1938 Ulster Grand Prix, Harold Hartley succeeded in pushing his 250 Rudge into third place in the 250cc Class by half distance, but retired on the next lap. It was a stout effort though. Chris Tattersall then lying fifth at that stage took his Python-engined (Rudge-engined) CTS into third spot at the finish, at an average speed of 74.70 mph (120.29 km/h) some 2½ minutes behind the winner Ernie Thomas on the supercharged DKW.

In June 1938 development work started at the Hayes factory on a new WD machine based on the 250 Rapid and at the end of the year it was submitted to the War Department for approval tests as a potential despatch rider's machine. This was fitted with a 500 gearbox, which was considered to be more durable under the rough treatment expected to be meted out to such machines during wartime conditions, but the 250 type of gearbox adjuster was retained. A heavier clutch with four friction discs was employed and to meet Government specifications an offside gearchange lever was adopted using 500cc machine parts. Standard 250cc machine brake shoes were used but the nearside brake drum and bolted-on rear sprocket were again 500cc components. Once again, to meet WD requirements the central stand was discarded in favour of a rear stand and a prop stand. A Lucas Magdyno provided both the ignition and lighting, and proved vastly superior to the Maglita used previously.

The War Department eventually accepted the 250 WD Rudge and the first batch of 200 machines was delivered. Later, years after the cessation of hostilities, some of these models were sold off to the general public and their combination of 500cc and 250cc components created the impression that they were in fact 'bitzas'. In fact few actually survived, most being destroyed when the Hayes factory was bombed early in 1940.

Before leaving 1938, it is worth mentioning a most unusual racing car application for the 499cc Rudge engine, that took place that year. Richard Bolster, the brother of the late John Bolster, was then racing a GN fitted with an 1100cc supercharged MG engine. It was raced at Donington and at Shelsley Walsh, but never proved really competitive. When it eventually decided to throw a connecting rod through the side of the engine block, Richard replaced the unit with four dirt-track Rudge engines of around 1931 vintage. These were placed with their crankshafts longitudinally down the centre of the chassis one behind the other. A shaft ran down inside the offside chassis member, supported in ball races carried in housings on the engine-bearer arms.

This shaft was connected by chain to each of the engines and a further Duplex chain drove a flywheel and Borg & Beck clutch. The latter was supported on a cross member in the centre of the chassis and drove the normal GN propeller shaft. It should have been a very lively 2-litre car, but there were snags. As John Bolster later recalled: 'In practice, the car was just as fast as one would expect, but it was extremely difficult to start without breaking some part of the inter-engine transmission. If all four started instantly, all was well, but the slightest backfiring or snatching invariably broke something.'

The 1939 Lightweight TT results proved fairly encouraging for the company, with Harold Hartley finishing in seventh place. Rex Judd had entered the two Pike brothers and Ginger Wood in the race as a

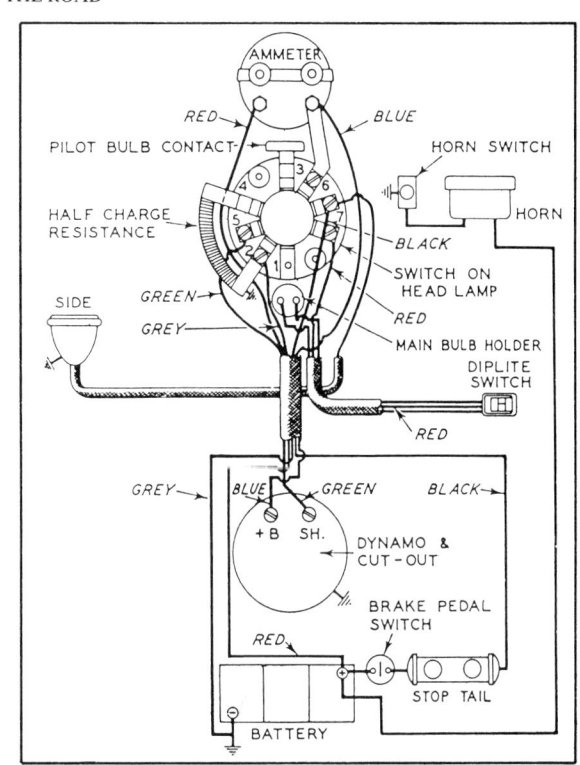

The wiring diagram of the Miller electrical equipment used on the later Rudge-Whitworth motorcycles.

Rudge team. Wood came ninth, Roland Pike eleventh, while his brother retired with engine trouble.

In the 250cc Class of the Ulster Grand Prix Chris Tattersall lay second on his Rudge-engined CTS with A. Glendinning (Rudge) joint third with Hulme (Excelsior) at the half-way stage. Glendinning retired on the ninth lap, but Tattersall managed to finish third at the fine average speed of 74.62 mph (120.16 km/h).

With the outbreak of World War 2, the Hayes factory was well prepared for a large production run of WD 250 machines after the initial batch. But this was not to be, for the unforeseen was to change all that. The Rudge-Whitworth factory was immediately adjacent to the electrical equipment production plant and was an obvious choice for any future expansion of the latter with the increased demand for radar and its supporting equipment at the climax of the Battle of Britain in 1940. At that time the Gramophone Company (now better known as HMV) needed all the space it could get to quickly produce this equipment. Whether or not the story that went round that Lord Beaverbrook, then Chief of Aircraft Production, during his visit to the plant had ordered its closure and switching over to the production of urgently needed radar equipment, be true or not is largely irrelevant since the change would almost certainly have happened anway in such a situation. What is sad is that production of motorcycles did not carry on after the war, for during 1939 by the time the war had been declared, motorcycle production for the year had already topped the 2000 mark and was still rising rapidly and might well have reached a grand total of 4000 machines by the end of the year.

Permission was granted, however, for The Norman Cycle Company of Ashford in Kent, to continue production of the autocycle using the Rudge name and it was to be the last production, powered two-wheeler to be so named.

Even at this early stage the firm already had in hand a range of machines for 1940, which would have comprised the two-valve 250 Rapid and the two-valve 250 Sports model, the 250 two-valve WD machine, the 500 Ulster and 500 Sports Special and a quite new design of engine for incorporation in the 350 machine. Two types of power unit had been designed, both with two valve heads, the first a pushrod-operated unit the second with an overhead camshaft. The pushrod version was not dissimilar to the high-camshaft Velocette engine. The war put paid to all this development though and, as is well known, EMI did not see fit to produce Rudge-Whitworth motorcycles after World War 2. This latter decision was made in 1943 when the old Rudge concern was sold off, the trade mark (and rights to produce bicycles) passing to Raleigh Industries that year and used for a few of that group's post-war cycles.

15
Old Rudges never die (1946–1985)

After World War 2 Godfrey's of Great Portland Street, London, who had acquired the right to produce spares for Rudge-Whitworth machines, looked after the needs of the remaining Rudge-riding motorcyclists. Godfrey's in turn obtained spare parts from subcontractors, notably A.L. Bailey of Walthamstow, who had been a Rudge-Whitworth dealer before the war. During the period of hostilities he had converted his showroom into a machine shop and so was well set up to produce Rudge parts when peace returned. By 1950, Mr Bailey had designed and built what he considered to be an up-to-date version of the Rudge. It was the post-war period when all the major factories were producing vertical twins. Bailey had mounted two Rudge-Ulster heads and barrels on a common crankcase to form a V-twin of 998cc. Mounted in a hydraulically damped, swinging-arm-spring frame of his own design and with Matchless Teledraulic front forks, it was strictly a one-off prototype. Nevertheless his aim was to eventually produce a range of machines including 250, 350, 500 and 1000cc models, all under the Rudge name and he actually got to the stage of producing an all-aluminium alloy 250 single-cylinder power unit. These ambitious plans fell down, due to lack of finance, as is so often the case, and that was the end of that particular project.

On the road racing front, both Harold Hartley and the Pike brothers rode again after the war. Harold achieved an eleventh place in the 1949 Lightweight TT Race on his 250 Rudge, tenth in the 1950 race and 17th in 1952. The latter result was really quite good for a machine well over 15 years old at the time.

The Pike brothers went a lot better after the war, having modified their machines to incorporate all the latest developments. Stan Pike finished fifth in the 1947 Lightweight Race at an average speed of 68.56 mph (110.40 km/h) on his much modified Pike-Rudge. It must be remembered that in those days racing was carried out on the notorious engine-wrecking fuel known as Pool Petrol having an octane rating of around 72 — hence the much lower speeds compared with the pre-war era when petrol-benzole was available.

In the 1948 Lightweight TT event, his brother, Roland Pike, finished second to Maurice Cann's Moto Guzzi, at an average speed of 71.86 mph (115.92 km/h). In the 1949 race, Roland finished third at 72.94 mph (117.46 km/h) with his brother Stan seventh at 69.08 mph (111.24 km/h) and Harold Hartley eleventh at 68.04 mph (109.56 km/h). The 1950 Lightweight TT Race saw Roland Pike's average speed rise to 74.14 mph (119.39 km/h), but he could do no better than fourth spot.

During the late 1940s and in the 1950s when 500cc car racing was highly popular, most constructors resorted to using motorcycle power units and some of them were of Rudge origin.

One of these cars, built by Frank H. Bacon and christened the FHB 500, employed a 1931 Austin Seven chassis and a 1938 500 Rudge-Ulster engine. The original intention was to couple up the power unit direct to the Austin transmission, but the crankcase proved too wide to go between the frame members. It was therefore decided to mount the engine above the frame and drive the clutch by a short vertical chain. The lightened Austin flywheel, which carried the clutch, was mounted on ball races supported from the chassis.

The chain proved far too short for reliability and so was lengthened and carried around a jockey pulley. The car ran in this form in 1947 and was then converted to the more usual rear-engined layout. In this form the power was transmitted via a four-speed Burman gearbox and drove the rear axle by chain. A neat body completed the car.

Yet another 500cc racing car employing a Rudge Ulster engine was the GSI built by Gerald Spink, which was nicknamed by other drivers the 'Squanderbug'. This had a light chassis made from 14-gauge steel channel and its engine was located at the rear on a three-point rubber mounting. The primary chain drove a Norton four-speed gearbox and from this the drive was taken by a short chain to a short shaft having supporting races in housings on the chassis. Other details included: the use of hubs, races and housings comprising the complete ends of a Wolseley Hornet back axle, while the front suspension was from a Morgan three-wheeler.

Godfrey's continued the production of Rudge spares right up into the early 1960s, but when the firm decided to close down its Great Portland Street depot and operate solely from Croydon, the manufacture of Rudge components ceased.

One might have thought that this would have been the end of the Rudge-Whitworth story — but no. With the formation of the Rudge Enthusiasts Club, following the disbandment of the Rudge Association formed after World War 2, interest in the marque has been kept alive. Even more amazingly some of these admittedly classic machines are now performing amazingly well in so-called drag racing — or what used to be known as sprinting.

One of the exponents of the 250 Rudge in sprints in the period 1955 to 1960 in events organised by the National Sprint Association, was the slightly built and bearded Dave Tringham. I well remember seeing him in action winning the 250cc Class at Ramsgate and other sprint venues on his 250 four-valve machine. Amazingly he had modified this single cylinder power unit to accept twin carburettors, but how he managed to get the mixture settings right has always baffled me.

More recently during the late 1960s, Bill Orriss and D. Leigh, riding 349cc four-valve Rudges, basically at least of 1933 vintage, have put up some staggering sprint performances. One of the most remarkable of these took place back in the late sixties at the Vagabond Club's standing-start quarter-mile dash at Duxford near Cambridge, at the end of July 1967, when they came first and second in the 350cc Class. Bill Orriss won with an amazing time of 11.40 sec, while Leigh came second in 12.46 sec. Since then Orriss has, I believe, broken the 11 sec barrier and set a number of sprint records. As you can see, old Rudges never die, they simply fade away like old soldiers.

Appendix 1

Some representative Rudge-Whitworth valve and ignition timings

Model	Year	Inlet Opens BTDC	Inlet Closes ABDC	Exhaust Opens BBDC	Exhaust Closes ATDC	Ignition advance BTDC	Tappet clearances (in) Checking In	Checking Ex	Running In	Running Ex	Comments
3½ hp ioe single	1911–19	0°	24°	61°	20°	2.5mm	0.031	—	nil	0.010	M370×6 overlap cam
	1911-19	10° ATDC	22°	53.5°	0°	2.5mm ATDC	0.031	—	nil	0.010	M558×1 std cam
	1920-22	0°	24°	63.5°	0°	2.5mm ATDC	0.031	—	nil	0.010	M506×10 cam
	1923	23.4°	47.4°	57.3°	17.3°	2.5mm ATDC	0.031	—	nil	0.010	M1459×1 cam
5-6 hp ioe single	1913-19	5° ATDC	30°	52°	16°	3.0mm ATDC	0.031	—	nil	0.010	C94×7 std cam
	1920-21	0°	24°	63.5°	0°	3.0mm ATDC	0.031	—	nil	0.010	M506×10 cam
7-9 hp ioe twin	1915-23	0°	31.5°	60°	0°	4.0mm ATDC	0.031	—	nil	0.010	T127 1 & 3 cam
500 ohv Special /Sports Special	1928-39	0.2mm	8.4mm	14.4mm	2.4mm	12-14 mm on full adv	0.020	0.020	nil	0.004	—
500 ohv Replica	1931-32	9mm	14.4mm	14.4mm	9mm	18mm on full adv	0.010	0.010	nil	0.003	—
350 ohv Replica	1931-33	9mm	14.4mm	14.4mm	9mm	16mm on full adv	0.008	0.008	nil	0.003	—
500 ohv Ulster	1933-39	10mm	13mm	16mm	10mm	12-14mm on full adv	0.020	0.020	nil	0.004	—
250 ohv 4-valve Sports	1936	8mm	13.2mm	13.2mm	8mm		0.010	0.010	nil	0.004	—
250 ohv Rapid	1936	0.2mm	8mm	13.4mm	2.2mm	12-14mm on full adv	0.025	0.025	nil	0.004	—

Appendix 2

The last frame numbers in any given year on Rudge-Whitworth motorcycles

Table 1: 1910-1914 models inclusive

	3½ hp ioe single							
	Single gear					Variable gear		
Year	Roadster (Fixed gear)	Free engine (Clutch)	TT (Fixed gear)	Brooklands (Fixed gear)	3-speed (Hub gear)	Roadster Multi	TT Multi	5-6 hp ioe Multi
1910	592203	—	—	—	—	—	—	—
1911	635225	635300	635088	—	—	—	—	—
1912	670451	670973	670970	669408	—	670975	—	—
1913	705902	705943	705592	703064	—	705942	—	705923
1914	731028	731992	732310	727696	732193	732317	732311	732086

Table 2: 1915-1923 models inclusive

	Fixed gear Brooklands model	3½ hp ioe single						5-6 hp ioe single			7-9 hp ioe twin		
		Variable-gear models											
		3-speed hub		Multi gear		Gearbox							
Year		Roadster	TT	Roadster	TT	3-speed	4-speed	Multi gear	3-speed gearbox	3-speed hub	Multi gear	3-speed gearbox	4-speed gearbox
1915	740937	738126	738424	742695	742658	—	—	741645	—	—	738742	—	739200
1916	—	—	—	751301	—	—	—	—	—	—	—	—	—
1917	No machines produced by Government decree												
1918	—	—	—	—	751325	—	—	746577	—	—	745476	—	—
1919	—	—	—	759966	760306	—	—	760270	—	—	760649	—	—
1920	—	—	—	774899	776004	—	—	775478	—	—	775479	775702	—
1921	—	—	—	790121	794428	794209	—	787831	—	—	788351	794208	—
1922	—	—	—	813255	814937	814683	—	—	—	—	—	813481	—
1923	—	—	—	829812	833273	—	837398	—	—	—	—	—	833476

Table 3: 1924-1930 models inclusive

	3½ hp ioe 4-speed single	350 ohv 4-valve	500 ohv 4-valve				
Year			Standard	Sports	DT	Special	Ulster
1924	839288	2800*	2800*	—	—	—	—
1925	—	12000*	12000*	12000*	—	—	—
1926	—	—	20000	20000	—	—	—
1927	—	—	24605	—	—	—	—
1928	—	—	—	—	—	31120	—
1929	—	38418	—	—	—	—	—
1930	—	—	—	—	—	45200	—

* Approximate numbers, since records were destroyed.

Table 4: 1931-1939 models inclusive

Year	4-valve Sports	250 ohv 2-valve Replica Sports	Tourist	Sports/ Rapid	500 ohv 4-valve Special	Ulster
1931	—	—	—	—	47214	—
1932	—	—	—	—	—	50422
1933	—	—	—	—	52369	—
1934	54604	—	—	—	—	—
1935	—	—	56264	—	—	—
1936	—	—	—	58980	—	—
1937	—	—	—	62861	—	—
1938	—	—	—	64752	—	—
1939*	—	—	—	—	—	—

*Last frame number for a 1939 production machine was 65720.

Index

Abbott, Ray ('Milky'), 36-8
ABC motorcycle, 54
Acetylene system, 52
Adamson, J. W., 42, 44-5
Admiralty, The, 49
AJS motorcycles, 39, 72, 82-4, 88-9, 104
AJW motorcycles, 97
Alan Trophy Trial, 80
Alcohol fuel, 63
Alexander, 'Alfie', 36-7
Alkmaar, 75
Aluminium-alloy head, 115; aluminium-bronze head, 106, 115
Ametila Circuit, 75
Ammunition, production of, 49
Amott, Jack, 66-8, 71
Amsterdam Grass Track, 68
Anderson, Fergus, 103, 108
Anstey, Frank, 98
Aprica Hill Climb, 34
Aquila motorcycles, 97
Ariel, cycle factory, 13; 'high' bicycle, 6; motorcycles, 106; spring frame, 98
Army Ordnance Corps, 47; Service Corps, 47
Arthur, Frank, 79
Athy Road Race, 66
Austin '7' chassis, 121
Australian National Record, 75, 80; TT Races, 91
Austrian TT Race, 81, 101
Autocycle, 120
Auto Cycle Union (ACU), 21, 23, 31, 51, 82, 84, 112; doctor, 44; North-West Centre Championship Trial, 68; Quarterly Trials, 19; Round Midlands Tour, 17; Senior One-Hour (1913) Brooklands' Championship, 40

Bacon, Frank H., 121; H. J., 103
Bailey, A. L., 121
Baldwin, O. M., 52-3, 55
Ball-bearing factory, 49
Ballot racing car, 80
Bandini, Terzo, 75
Barrington, Manliff, 108
Barwell, 84
Bateman, Frank, 33-7
Bath & Avon Rose Bowl, 80
BAT-JAP motorcycles, 19-20
Battle of Britain, 119
Bayly, C., 110
Beach, H. H., 56
Beaverbrook, Lord, 119
Belgian Army, 48-9; Belgian Grand Prix, 64, 95-6, 101, 105, 108
Bemrose Trial, 99
Bennett, Alec, 73
Benzole Handicap, 38
Best & Lloyd pump, 53, 62-3, 66

'Bicyclette' safety cycle, 9-11
Birmingham Club, 60, 79, 91
Bluebird, 75
Bolster, John, 119; Richard, 119
Bolton, D. C., 29-30
Borg & Beck clutch, 119
Borgo, Alberto, 58; Carlo, 58
Bosch magnetos, 55
Bovy, Albert, 97; motorcycles, 97
Bowen, Harold, 20
BMW motorcycles, 106
Braham Cup, 79
Brakes, bi-metal drum, 106; constant clearance shoe, 78; coupled, 61; six-shoe, 63
Bradbury, Dan, 24
Brandon Speedway, 87
Brewin, Alf, 110
Brico rings, 86
'Bridle-rod' steering, 9
British Motor Cycle Racing Club (BMCRC), 19-21, 33, 39, 47, 96; 'Cup' Race Meeting, 68, 73, 89; 21st Anniversary Race Meeting, 80; 200-Mile Solo Races, 68, 73, 89
Bristol Cup, 80
Brittain, Vic, 78
Brooklands Automobile Racing Club (BARC), 20, 30, 34, 38-9, 45; Brooklands Gold Stars, 68, 96, 103, 106, 112; Grand Prix, 106; Test Hill, 18; Track, 14-15, 17, 19, 21-2, 27, 37-8, 41, 45, 51, 54, 78, 80, 89, 96, 101, 106, 108, 110, 114
Bruce, Alan, 75
Brussels Motorcycle Show, 58
Burney, Charles, 17-20, 24
Byrne, J. J., 92

C & H-JAP motorcycle, 40
Cadwell Park, 86
Campbell, Malcolm, 73
Cambrian Trial, 99
Cann, Maurice, 121
Canteloup, R. E., 59
Carburettors, Amac, 63-4; Amal Type TTA, 29, 86; bell mouths, 99; Brown & Barlow (B & B), 16-17, 22; downdraught, 107; Senspray, 28, 47, 61
Carvill, A., 19
CAV magnetos, 22, 50-51
Championship of Russia, 38
'Cheltenham flyer'; See Williams, Jack; Cheltenham Challenge Cup, 79
Chemical and Physical Laboratory, 13
Chevell, B. A., 106
Chirney, V., 103
Circuit de Picardie, 38
Citroën, 38

Clady Circuit, 67
Claremont Speedway, 79
Clark, F., 33
Clarke, Henry, 7-8
Clarke Challenge Cup, 79
Clipstone Speed Trials, 24
Cocker, Jimmy, 40-41
Cogent Cycle Company, 7
Coleman, Percy, 86
Collier, Charlie, 20, 40; Harry, 44
Collins, C., 75
Colver, Bert, 44
Cooper, Frank, 11
Cotswold Cup Trial, 79, 91
Cottle, Margaret, 79
Cotton, 97
Courtney, L. R., 96
Coventry Lever Tricycle, 9
Coventry Machinists Company, 7; Coventry Rotary Tricycle, 9-10
Cox, W., 112
Crabtree, Syd, 107
Crash helmets, 42; See also Safety helmets
Cremonese Circuit, 35
'Crescent' bicycle, 12
Crimea, 8
Crimean War, 7
'Crocodile', The, 10
Croft, Norman, 114
Croombs, T., 69
Cross rotary valve, 110-11; Cross, T. D., 110-11
Crossley ambulances, 49
Crow Lane, 9, 13
Crystal Palace Cycle Show, 13; Crystal Palace Track, 10
CTS motorcycles, 97, 101, 105, 113-14, 118-19
Cussens, H. S., 71
Cycle & Allied Trades Association, 13
Cycle Components, 12
Cycle Trader, The, 13
Czechoslovakian Grand Prix, 74

Daily Express, The, 38
Dalton, Jim, 72, 85, 91
Davenport, Leo, 88
Debenham, A. T. H., 103
Denly, Bert, 89
Denmark, King of, 49
Denney, T. W., 71
Despatch Department, 13
Dicker, Bob, 53-6, 67
Dickinson Cup, 88
Dinter, A. W. van, 75
DKW motorcycle, 118
Dodson, Charlie, 66-8, 72-3, 82-3, 88, 105, 107
Doglioli & Civardi, 97
'Double-Twelve' 24-hour record, 54
Douglas High School, 44
Driscoll Cup, 96, 106

Dublin & District Motor Cycle Club Hill Climb, 19
Dunelt motorcycles, 97
Dunfee, Jack, 80
Dunlop tyres, 22
Duralumin connecting rod, 95; pushrods, 106
Durandal motorcycles, 97
Dutch 1100-km Trial, 68; Dutch TT Races, 67, 74, 95, 101, 105
Duzmo motorcycle, 30

Eadie, Albert, 12
Earls Court Motorcycle Show, 114
'Easy Payments Scheme', 12
Economic depression, 89
Eden, Mount, 60
Edge, C., 71; Selwyn F., 11, 14
Edward, J. E., 112
Elce, Billy, 20, 22-3, 25, 30
Elektron alloy, 60, 94
EMI Group, 111, 115, 120
Enfield Cycle Company, 12
Escol motorcycles, 97
Essex Motor Club, 54
Estonian TT Races, 108
European Grand Prix, 87
Excelsior motorcycles, 107, 112; 'Mechanical Marvel', 105
Eysink motorcycles, 97, 107

Faura, 76
Ferdinand, Archduke Franz, 46
FHB 500 racing car, 121
FICM, 51, 71, 75, 95
FitzGerald Cup, 112
FN motorcycle, 84
Follis motorcycles, 97
Fossati, 35
Frame numbers, 124-5
Friedland Annual Reliability, Trial, 68
Frogley, Roger ('Buster'), 69
Fry, G. H., 45, 47
Fuel experiments, 97
Fumagalli, Carlo, 106

Galloway, Dr, 108
George VI, HM King, 54
German Grand Prix, 67, 74, 95, 108
Gibson, John, 20, 23
Giro d'Italia, 44
Glanfield, Stan, 64-5; Glanfield Lawrence Ltd, 69
Glasgow Exhibition, 56
Gleave, Sid, 105
Gledhill, Norman, 88
Glendinning, A., 119
Glen Helen Hotel, 23, 73
Glover, P. E., 33-4
GN car, 119
Godfrey's, 118, 121

126

INDEX

Good, L., *112*
Gotham Cup, *79*
Governor's Trophy, *96*
Graham, H. H., *104, 109*
Graham-Walker Syndicate, The, *104-9*
Gramophone Company, The, *111, 119*
Grand Prix d'Europe, *64, 67*
Grand Prix of Denmark, *38*
Gray, G. T., *22, 36*
Greek TT Races, *108*
Greene, Tommy, *38, 40-41, 43-5, 47*
Greenwood, A., *75*; J. H., *112*
Grindlay-Peerless motorcycle, *68, 73, 78, 97*
Grose, Gus, *80, 89*
Gschwilm, Georg, *78-9*
GSI racing car, *121*
Guthrie, Jimmy, *82-4, 92, 95*

Hack, George, *50, 52, 58, 60, 63, 66, 68, 73, 78, 81-5, 87, 89, 91-2, 94, 97, 108-9, 112, 115*
Halford, Major Frank, *57*
Hammersveld, A. W. van, *75, 96, 107*
Handley, Wal, *67, 72, 84-7, 93-4, 96, 99-102*
Harley-Davidson motorcycles, 'Peashooter' Harley, *69*
Harris, Ron, *110, 112*
Hartley, Harold, *112, 114-15, 118, 121*; Laurence, *51*
Haswell, Jack, *39*
Haynes & Jefferis, *9*
Heales, S., *33-4, 39*
Healey, Joseph, *19, 23*
Heathcote, H. L., *13*
Heenan & Froude brake, *87*
Hicks, Freddie, *82, 84, 88*
Hill, B. Alan, *19, 22*; H., *33*; L., *33-4, 40, 45*; L. P., *105, 109*
Himing, George, *87*
Hilversum Grass Track Races, *68*
HMV, *119*
Hoare, P. H. T., *40*
Holroyd, Victor, *12-13, 49-50*
Hooley, Terrah, *12*
Horsman, Vincent Edward ('Victor'), *33, 39*
Houghton, H., *23*
Howell, Richard, *10*
HRD motorcycles, *75, 97*
Hunt, Tim, *72-3, 75, 83, 92*
Husqvarna motorcycles, *107*
'Hutchinson Hundred' Race, *67, 112*
Huxam, J., *80*

Illustrated Encyclopaedia of Motorcycles, The, *58*
Imperia motorcycles, *97*
Indian motorcycles, *31, 35-7, 44, 56*
International Road Race (Italy), *34*
International Six-Days' Trial (ISDT), *71, 96-8, 108, 112, 114*
Ireland, J. B., *71*
Irish Road Racing Championships, 25-Mile, *67*
Irving, Captain, *89*
Isle of Man, Car TT Race (1908), *14*; 'Mountain Circuit', *23*; Motorcycle TT Race, Coronation Jubilee (1911), *23*; (1912), *31*; (1913), *34-7*; (1914), *42-3*; (1922), *52-3*; (1928), *66*; (1929), *72*; (1930), *80-7*; (1931), *92-5, 97*; (1932), *99-102*; (1933), *105*; (1934), *107-8*; (1935), *109*; (1936-9), *111, 113-14*; (1939), *119*; (1947), *121*; (1948), *121*; (1949), *121*
Italian Army, *49*; Grand Prix, *75, 105*; Motorcycle Show, *58*

James, Sammy, *85*
JAP engines, *31, 69, 77*
Jelpke, F. W., *54*
Jenness, W. J., *110*
Jock, the Scots mechanic, *83*
Johnston, Paddy, *96*
Judd, Rex, *89, 119*

Kemble, H. R., *78*
Kent, HRH Prince George of, *102*
Kickham Trial, *80, 99*
King's Bench Division, *13*
'King's Head' Hotel, *11*
Kitchen, Bill, *105*
Kolberg Race Meeting, *75*
Kynaston, G. H., *106*

Lacey, Bill, *54, 68, 78*
Ladies' Cup, *78*
Lady motorcycles, *97*
Lafone, Jack, *103, 106*
Lamb, *72*
Lario Circuit, *75*
Lawson, Harry, *9-10, 12*; first motorcycle designed by, *10*
Lea Bridge Speedway Track, *69*
Leeds-to-London Reliability Trial, *27*
Leeds £200 Trial, *79*
Leicester Super Speedway, *77, 86*
Leigh, D., *122*
Leinster '100' Road Race, *67, 101*
Lermitte, Miss Betty, *78-81, 96*
Le Vack, Bert; *See* Vack, Bert Le
Levene, Jack, *89*
Leyden, Jock, *107*
Linx motorcycles, *97*
Lloyd, Major Lindsay, *22*
London-to-Edinburgh Run, *24, 35*; London-to-Gloucester Trial, *78*; London-to-Land's End Run, *12, 62*
Longman, Frank, *63, 66*
Loweth, Lance, *96*; Loweth-JAP motorcycle, *96*
Loughborough, Tom, *31*
Lowndes, Matthew 'Jumbo', *10*
Lucas electric lighting, *52*; Magdyno, *109, 118*; Maglita, *109, 117-18*
Lycett Aero saddle, *77*
Lyso driving belt, *20, 22*

Mabon, A., *22*
M & C; *See* Marriott & Cooper
MacDonald, J. A. *103, 108*
McDonagh, A. J., *33-4, 39*
MacGregor, Bob, *73, 80, 96, 112*
MacLeslie, J. A., *112*
McMeekin, H., *45*; J., *43-4*
Mair, R. A., *112*
Manders, Charlie, *96, 101, 105, 112*
Manx Grand Prix, *88-9, 118*

March, Vernon, *38, 40*
Marchant, Dougal, *67*
Marriott & Cooper, *12*
Marshall, W. C., *96, 103*
Martin, Les, *105, 107-8*
Mary, HRH Queen, *55*
Mason, Hugh, *35*
Matchless motorcycle, *20, 35, 40, 44, 55*; Teledraulic forks, *121*
Mathers, Bert, *54-6*
Matthews, M. R., *68*; P. H. A., *37, 44-5*
Maund, Captain, *54*
Megaphone exhausts, *91*
Meldi motorcycles, *97*
Mellors, Ted, *93-5, 107*
Meredith, C., *38*
Merrill, Ralph, *88, 92*
Meyrin Circuit, *67*
Military staff cars, *49*
Miller electrical equipment, *119*; headlamp, *76*
Miller, Max, *56*
Miller-Balsamo motorcycles, *97, 106*
Minett, Dennis, *112, 114*
Ministry of Munitions, *49*
Miniature motorcycle and cycle, *55*
Mitchell, Alec, *101*; Ernie, *96*
ML magnetos, *51, 55, 62-3, 70, 76*
Molineux Gardens, *8*
Montlhéry Track, *60, 68, 78*
Morgan, Harry, *31*
Morgan three-wheeler, *29, 31*
Morley, Jack, *10*
Moto Borgo motorcycles, *58*; Cambio Graduale, *58*; Tipo Tourismo, *58*
Moto-Guzzi motorcycles, *103, 105-6, 109, 121*
Motopedale motorcycles, *97*
Motor Cycle, The, *16, 106*; Challenge Cup, *40*; Clubman's Day, *110*
Motor Cycling, *32, 78, 114*
Motor Cycling Club (MCC), *35*; London-Exeter Winter Run, *19*; Reliability Trial, *27*
Motosacoche motorcycle, *67*
Multi gear; *See* Variable gears

Napier aero engines, *30*; racing car, *14*
Natal Spruit Race Meeting, *80*
National Motor Museum, *67*; National Physical Laboratory, *73*; National Sprint Association, *122*
Nazis, *112*
Neisse, R., *71*
Neudenberg, *38*
New Henley motorcycle, *75*
New Hudson motorcycle, *68*
New Imperial motorcycle, *93, 105, 107*
Newman, Harold, *112*; Howard, *34*
Newsome, Billy, *20, 22*
New Rap motorcycles, *97*
Nixon, Alfred, *10*; Basil, *75*
'No racing' statement, *104*
Norman Cycle Company, The, *122*

Norton motorcycles, *24, 40, 54, 56, 72-3, 83, 87-8, 92-3, 96, 100, 105*; gearbox, *121*
North-West '200' Road Race, *72, 80-81, 91-2, 99, 105*
Nottingham Motor Club, *22*
Nott, G. E. ('Ernie'), *60-61, 63-4, 66-8, 71-4, 78-89, 91-6, 98-100, 102-4, 107-10*
NUT motorcycle, *35*

O'Donovan, Dan, *40, 74*
Oil pump, worm-driven (1929 TT), *71*
Ollerhead, Stan, *61-2*
Olympian Cycle & Motor Cycle Show, *17-18, 28, 52-5, 58-9, 70, 78, 89, 114*
'Olympia' tandem, *12*
Open Championship of Belgium, *38*
Orris, Bill, *122*
Ortueta, Javier de, *110*
Ouwerkek, Lambert van, *97*

Palmer Tyre Company Silver Cup, *22*
Parker, Jack, *87*
Pattison, B., *38*
Pennington motorcycle, *10*
Penya Rhin Circuit, *76*
Perry & Company, *12*
Phillips, Walter, *8, 12*
Phoenix Park, *5*
Pike, Roland, *116, 118-19, 121*; Stan, *118-19, 121*
Pilling, H., *101*
Pioneer Run, *79*
Pither, Frank, *31*
Polish TT Races, *68*
Pool petrol, *121*
Portland sidecar, *25*
Pountney, Frank, *31*
Povey, Fred, *71, 79-80*
Prendergast, J., *23-4*
Prestwich Cup, *96*; Prestwich, J. A., *89*
Pringle, Jimmy, *75*
Public Schools MCC, *37*
Puch motorcycle, supercharged racing, *95*
Pugh & Sons, *11*
Pugh, Bernard Vernon, *63*; Charles Henry, *11-12*; Charles Vernon, *12, 47, 49*; John Vernon, *12-14, 18, 20, 27, 30-31, 33, 38, 49, 51-2, 55, 57-8, 60, 63, 65, 69, 73, 78, 81, 84-5, 88, 90, 94, 97-8, 103-4, 112*
Pullin, A. L., *45*; Cyril, *38-45, 47*; designed frame, *44, 50, 53*
Python engines, *97, 105-6, 113-14*

Railway Workshops Wolverhampton, *12*
Rake Haw Cup, *80*
Raleigh Industries, *120*; Raleigh motorcycles, *67, 74, 84*
Ransom, R. P., *116*
Ransome & Marle (R & W), *104, 106*
Rea Street, *11, 49*
Record attempts, *20-2, 24-7, 30, 41, 54, 56, 60, 68, 78, 97, 110*

127

Refoy, P. A., *108*
Rex Acme motorcycles, *97*
Reynold, G. L., *96*
Rhys, W. L. T., *25-6*
Ricardo, Sir Harry, *57*; 'Ricardo Triumph' motorcycle, *57*
Rigg, W. H., *96, 103*
'Road Sculler', *11*
Round crankcase flange, *71*
Rowlandson, F. A. ('Rowley'), *34-5, 38-9, 44-5*
Royal Grand Prix of Rome, *96*; Royal Naval Air Service, *47*
Rudge Association, *122*
Rudge Book of the Road, The, *64*
Rudge caravan, *65, 79*
Rudge Enthusiasts Club, The, *80, 122*
Rudge, D., & The Coventry Tricycle Company, *9-10*
Rudge, Harry, *9, 12*
Rudge Cycle Co. Ltd., The, *12*
Rudge 'Ordinary', *10*; Quadricycle, *11*; Record, The, *10-11*; 'Rotary', The, *14*; Tandem, *12*
Rudge-Whitworth, Cycle Club, *14*; depot system, *14*
Production models:
Brooklands, *28*
Fixed-gear, *18, 48*
Free-engine, *18, 48, 77*
TT, *19, 28, 30*
Multi, 3½hp, *28, 30, 50, 52-3, 58*; 5-6hp, *52*
TT Multi, *44, 48, 52-3*
TT Roadster, *48, 50*
Multwin, 7-9hp, *48, 52*
'Colonial', 3½hp, *32*
Three-speed, hub-gear, 3½hp, *46*; 5-6hp, *46*
Three-speed gearbox, 3½hp, *53, 57*; 7-9hp, *53*
Four-speed, gearbox, 3½hp, *57*; 7-9hp, *53*
Four-valve, parallel, 350cc, *58, 60*; 500cc, *58, 60*; 500cc Export Racer, *63*; 500cc Sports, *65*; 500cc Standard, *65, 90*; 500cc Special, *65, 76-7, 90, 104, 114*; 500cc Dirt Track Special (DT), *70, 90*; 500cc Sports Special, *114, 120*; 500cc Ulster, *70, 76, 114*; 500cc Ulster-Replica, *90*
Four-valve, fully-radial, 250cc Standard, *106*; 250cc TT Replica, *98, 104-6*; 250cc Sports, *106, 109, 111*; 250cc Tourist, *109*; 350cc Standard, *76, 90*; 350cc TT Replica, *90*; 500cc Ulster, *97*; 500cc Ulster-Replica, *97*
Four-valve semi-radial, 500cc Ulster, *114, 120*
Two-valve, ohv 250cc Tourist, *109*; 250cc Sport, *115, 120*; 250cc Rapid, *111, 120*; 250cc

JAP, *69, 77, 89-90*
Two-valve, sv 250cc JAP, *69, 77, 89-90*; 300cc JAP, *77, 89-90*
Two-stroke, 98cc Autocycle, *115*
Prototypes and experimental models:
175cc four-valve, *101*
250cc high-cam single, *120*
250cc Rudge-Villiers, *69*
250cc supercharged twin, *91*
7-9hp Dicker/Mathers records' machine, *58*
1,700cc in-line four for records, *51*
four-wheeled cyclecars, *31-2*
Pullin's TT winner, the details, *44*
'V'-four' records' engine, *89*
Sidecars, *33, 48, 60, 77, 90, 103*
Rudge-Whitworth wheels, *14, 49*
Rusk, Walter, *105*
Russian Army, *47-8*

Safety helmets, *42*
Sanderson, C. R., *71, 80*
Sandri, *106*
Sappemeir Grass Track Races, *68*
Sarkis, Joe, *79-80*
Science Museum, *49*
Scott, *34*; Scott motorcycles, *35, 44*
Scottish Exhibition, *58*; Scottish Six Days' Trial, *71, 80-81, 96, 98*
Second International Motorcycle Trial, *38*
Segura, *76*
Sheard, Tom, *44-5*
Sheffield & Hallamshire MCC, *24*
Sherborne, River, *13, 49*
Silencer, 'Brooklands-type', *77*; 'Fishtail-type', *111*; Burgess-type, *105*
Simcock, Arthur ('Digger'), *74, 87, 101*
Simpson, Jimmy, *66, 92-3, 100, 107-8*
Singer motorcycle, *20, 29-30, 33-4, 38-9*
Six-Hour Multi-Class Scratch Race, *39*
Skilton, *14*
Sleightholme, J. T., *71*
Smith, Bob Walker, *11-12*; H. G. Tyrell, *66-8, 71-4*; *79, 81-2, 84-7, 92-5, 99-101, 103-5, 108-10, 112*
Smith, Starley & Co Ltd., *9*
Soresby, N. O., *30, 39*
South African Continental Cup, *38*; South African National Record, *75*; South African TT Races, *79-80*
Spanish Championship Races, *76*; Spanish TT Races, *101*
Sparkes, Lieutenant, S. W., *65*

Spencer, W. Stanhope, *22-7, 30*
Spink, Gerald, *121*
Splitdorf magnetos, *51*
Spon Street, *9, 12-13*
Sproston, A. J., *24*
'Squander Bug'; See GSI racing car
Staniland Trophy, *108*
Stanley, George, *20, 29-30, 33-4, 38-9*
Stealing racing parts, *104*
Stein, Sir Edward De, *106*
Steinfellner, *81*
Stewart, A. D., *96*; Gwenda, *60*; Colonel Neil, *60*
Straight, Whitney, *20*
Sunbeam motorcycles, *62, 65-6, 72-3, 78, 82-3, 87-8*; Sunbeam record attempt car, *89*
Surridge, Victor, *17-20, 23-4*
Swedish Grand Prix, *96*
Swiftsure, *9*
Syndicate, The; See Graham Walker Syndicate, The

Tangent & Coventry Tricycle Co Ltd., The, *9*; Tangent Bicycle, *9*
Tank, bulbous-nosed saddle, *61*
Targo Florio, *108*
Tattershall, Chris, *97, 113-14, 118-19*
Taylor, Colin, *105*; Vernon, *23, 33*
Taylour, Fay, *69*
Tecalemit grease gun, *77*
Temple, Claude, *67*; Temple Press Ltd, *114*; Temple Street Works, *7*
Tenni, Omobono, *109*
Tennigkeit, E., *75*
Territorial Army Signals Team, *112*
Tessier, Sidney, *19*
Thomas, Ernie, *118*
Tidswell, J. F., *14*
'Tiger's Head', *7-8, 11*
Times, The, *8*
Tooth, L., *112*
Townsend & Company, George, *12*
Trafalgar Street, *13*
Tringham, Dave, *122*
'Triple Eight' 24-hour record, *54*

Ulster Grand Prix, *66-8, 74, 87, 90, 96, 98, 101, 105, 108, 118*
Vack, Bert Le, *55-6, 65*
Vagabond MCC, *122*
Vailati, *34-5*
Valve and ignition timings, *123*
van Dinter, A. W.; See Dinter, A. W. van
Variable valve timing, *78*
Variable gears and gearboxes: Armstrong-Triplex, *24*; Jardine four-speed, *48, 52*

Mabon, *24, 28*
Multi, *27, 32, 37*
NSU two-speed, *24*
Rudge four-speed, *55*; five-speed, *78*
Sturmey-Archer, *46*
Zenith-Gradua, *24*
Varzi, *62*
Velandini, *47, 58*
Velocette, *82*
Velocipede, *8*
Verneuk salt pan, *75*
'Victor, E. H.'; See Horsman, 'Victor'
Victory Cup Trial, *60, 79, 91*
Villiers engines, *69*

Walker, Graham, *65-8, 71-2, 74-5, 77, 79-88, 91-2, 94-5, 98-101, 104-8, 114*; Murray, *101*
Wallace, John, *29*
Wallis, J., *12*; Wallis motorcycles, *79*
Walls, Patrick, *81, 96*
Ware, Edward Bradford ('Teddy'), *26, 29, 40*
Wassel, Jimmy, *80*
Watkinson, *89*
Weatherell motorcycle, *54*
Wedge, Charles Arthur, *12*
Werner Frere Company, *13*; Werner Motors Ltd, *13*
Wessex scramble, *77*
West, Jock, *106*
Whalley, Jim, *85*
Wheel building, *14*; Wheel hubs, ball-bearing, *8*
Wheeler, Rueb, *80*
White Trophy, *91*
Whitworth Cycle Company, *11-12*; Whitworth, Frank, Ltd, *61*; Whitworth jun., Frank, *61-2*
Wilcock, A., *68*
Williams, C. J., *74, 84*; Williams Cup, Eric, *79*; Williams, Cyril, *39*; Jack, *71, 78-80, 96, 109*; Stuart, *96*
Willis, Harold, *82*
Wills, 'Goldflake', *85-91*; P. L. B., *79*
Wolseley, *49*; 'Hornet' back axle, *121*
Wood, H. O. ('Tim'), *35, 37, 44*; G. W., *88, 92*; 'Ginger', *119*; Reg, *86*
Woodcock, George, *8-11*; 'Woodcock', The, *10*; Woodcock & Twist, *9*
Wright, F., *17*
Wye Valley Trial, *40*

Yano, K., *22*
York, HRH Duke of, *54-5*
Ypres, Battle of, *49*

Zenith motorcycle, *22*
Zundapp motorcycle, *97*